Translational Systems Sciences

Volume 7

In 1956, Kenneth Boulding explained the concept of General Systems Theory as a *skeleton of science*. He describes that it hopes to develop something like a "spectrum" of theories—a system of systems which may perform the function of a "gestalt" in theoretical construction. Such "gestalts" in special fields have been of great value in directing research towards the gaps which they reveal.

There were, at that time, other important conceptual frameworks and theories, such as cybernetics. Additional theories and applications developed later, including synergetics, cognitive science, complex adaptive systems, and many others. Some focused on principles within specific domains of knowledge and others crossed areas of knowledge and practice, along the spectrum described by Boulding.

Also in 1956, the Society for General Systems Research (now the International Society for the Systems Sciences) was founded. One of the concerns of the founders, even then, was the state of the human condition, and what science could do about it.

The present Translational Systems Sciences book series aims at cultivating a new frontier of systems sciences for contributing to the need for practical applications that benefit people.

The concept of translational research originally comes from medical science for enhancing human health and well-being. Translational medical research is often labeled as "Bench to Bedside." It places emphasis on translating the findings in basic research (at bench) more quickly and efficiently into medical practice (at bedside). At the same time, needs and demands from practice drive the development of new and innovative ideas and concepts. In this tightly coupled process it is essential to remove barriers to multi-disciplinary collaboration.

The present series attempts to bridge and integrate basic research founded in systems concepts, logic, theories and models with systems practices and methodologies, into a process of systems research. Since both bench and bedside involve diverse stakeholder groups, including researchers, practitioners and users, translational systems science works to create common platforms for language to activate the "bench to bedside" cycle.

In order to create a resilient and sustainable society in the twenty-first century, we unquestionably need open social innovation through which we create new social values, and realize them in society by connecting diverse ideas and developing new solutions. We assume three types of social values, namely: (1) values relevant to social infrastructure such as safety, security, and amenity; (2) values created by innovation in business, economics, and management practices; and, (3) values necessary for community sustainability brought about by conflict resolution and consensus building.

The series will first approach these social values from a systems science perspective by drawing on a range of disciplines in trans-disciplinary and cross-cultural ways. They may include social systems theory, sociology, business administration, management information science, organization science, computational mathematical organization theory, economics, evolutionary economics, international political science, jurisprudence, policy science, socio-information studies, cognitive science, artificial intelligence, complex adaptive systems theory, philosophy of science, and other related disciplines. In addition, this series will promote translational systems science as a means of scientific research that facilitates the translation of findings from basic science to practical applications, and vice versa.

We believe that this book series should advance a new frontier in systems sciences by presenting theoretical and conceptual frameworks, as well as theories for design and application, for twenty-first-century socioeconomic systems in a translational and trans-disciplinary context.

More information about this series at http://www.springer.com/series/11213

Shigeo Atsuji

Unsafety

Disaster Management, Organizational Accidents, and Crisis Sciences for Sustainability

Springer

Shigeo Atsuji
Kansai University
Kyoto, Japan

ISSN 2197-8832 ISSN 2197-8840 (electronic)
Translational Systems Sciences
ISBN 978-4-431-55922-1 ISBN 978-4-431-55924-5 (eBook)
DOI 10.1007/978-4-431-55924-5

Library of Congress Control Number: 2016938790

Printed on acid-free paper

This Springer imprint is published by Springer Nature
The registered company is Springer Japan KK

Preface: Unsafe Eden

Today, the conditions foreseen by Lester Brown, Jared Diamond, James Reason, and Michael Sandel are beginning to become a reality and manifest themselves in the immediate social environment and at a national and global level. Starting out from densely populated countries and spreading in all directions across national boundaries, the exploding ten billion population of the earth—Terra—(United Nations estimated figure for 2050) is progressively depleting food and water and energy resources for agriculture and industry. Humankind, like a huge swarm of locusts, is devouring resources and driving the whole Gaia system into a state of apoptosis.

Everywhere in contemporary society, the collapse, not only of weakened physical structures such as bridges, roads, and buildings but also of social systems such as pensions, nursing care, and health and social insurance, has drawn innocent people into unforeseen accident and disaster situations, causing great misery. This book presents examples of combinations of these human-made and natural disasters that developed into catastrophes. The book asks the readers what we need to do now to preserve Eden for our descendants, right down to the seventh generation. Appealing from Japan, a land of natural disaster in the Far East, I want people all over the world to think over this message of life and death.

Natural and human-made disasters happen worldwide and cause misery through loss of life; destruction of agriculture, fisheries, and other sources of livelihood; and interruption of urban life. Unsafety from a disaster in one place increases uncertainty elsewhere. Disaster can lead to famine, increased international tensions, refugee flows, and even war or revolution. In recent years, natural disasters have occurred frequently across the globe, while human-made organizational accidents have also followed an inexorable trend toward increase in scale, presenting urgent issues in all nations for individuals, organizations, regions, states, and the globe.

Unsafety focuses on the kinds of unnatural disaster and organizational accident which arise as repercussions of natural hazards, for example, in the author's native Japan, where earthquakes, tsunamis, and typhoons are common, with the Fukushima nuclear disaster as an outstanding example of this link between natural disaster and organizational accident. The author explain that one factor in the Fukushima catastrophe, which followed in the wake of the earthquake and tsunami, was the latent deterioration and aging of systems at all levels from the physical to the social, leading through chain reaction to unsought and unforeseen consequences. Here, the aging of the nuclear reactor system, the breakdown of safety management, and inappropriate instructions from the regulatory authorities combined to create a threefold disaster, in which technological, organizational, and governmental dysfunction have been diagnosed as reflecting a systems pathology infecting all levels.

This book examines accidents and disasters in the modern era and clarifies the mechanisms involved and the significance of emerging problems, from the aging of vital infrastructures for the supply of water, gas, and electricity to the breakdown of pensions, healthcare, and other social systems, demonstrating how we might check the underlying pathology and threat of systemic breakdown and exploring potential management systems approaches and policies bearing both on causes and effects.

The International Society for the Systems Sciences was founded by Ludwig von Bertalanffy and is associated with names such as Anatol Rapoport, Ross Ashby, Kenneth Boulding, Peter B. Checkland, Hal Linstone, Stafford Beer, Russell L. Ackoff, J.G. Miller, Ervin Laszlo, Howard T. Odum, and Ilya Prigogine. At its millennium world congress (2000), I was in immediate attendance at lectures by Humberto Maturana on 'autopoiesis' and Eric Chaisson on 'cosmic evolution', which left a great impression on me. Behind that emotion was a shared academic hinterland of classic works such as N. Wiener's *Cybernetics*, C.I. Barnard's *The Functions of the Executive*, H.A. Simon's *Administrative Behavior*, and the work of P.F. Drucker, R.L. Carson, T. Colborn, L. Silber, Kitarō Nishida, Wang Yangming, and Tetsurō Watsuji.

Especially, I appreciate their exploration of our future possibilities and am always inspired by the intelligent work of other academics, among them Lester R. Brown (Environmental Issues: Earth Policy Institute), Jared M. Diamond (Civilization Collapse: University of California), James T. Reason (Organizational Accidents: University of Manchester), Michael J. Sandel (Philosophy: Harvard University), Simon Bennett (Disaster Management: University of Leicester), Michael Morley who is the president of IFSAM (Human Resource Management: University of Limerick, Ireland), and my mentor Prof. Gerhard Chroust (Human-Made Disaster: Johannes Kepler University Linz, Austria). It was the inspiration from their work that led me to write the book, in which I quote from their important texts. I am grateful to their intellectual contributions for the rich inspiration and insight which they have given me.

Acknowledgments

In publishing this book, I wish to extend my heartfelt thanks for the valuable advice I received from Prof. Gerhard Chroust (Austria), Prof. Jennifer Wilby (United Kingdom), and Prof. Len Troncale (United States) while making presentations and giving some papers at the World Conference of the International Society for the Systems Sciences in the United Kingdom (2011), Vietnam (2013), and Berlin (2015), which were valuable study opportunities. Also, at the world congress of the International Federation of Scholarly Associations of Management in Berlin (2006) and Paris (2010) and when making a speech at its world congress in Ireland (2012) and Tokyo (2014), I was inspired by Prof. Michael Morley (Ireland) who is the president of IFSAM. Also, my colleague is Dr. Peiran Su of West Scotland University. In Japan, I am grateful for the encouragement of Prof. Shinichi Oota (Doshisha University), and Prof. Koichiro Hioki (Kyoto University). Grateful thanks also for the following endorsements:

> The book paints a distressing and eye-opening picture of the vicious conspiracy between large industry, governments and individual greed, sacrificing morality and sustainability for efficiency and profit. The compounding effect of human misjudgment, inadequate management, reckless face-saving and cheating, turns natural and human-caused disasters into multi-level catastrophes. Data, diagrams and maps of Japanese and world-wide catastrophes (Fukushima, etc.) support the statements. (Prof. Dr. Gerhard Chroust, Johannes Kepler University of Linz, Austria)

> While the genesis of Shigeo Atsuji's inspiration for *Unsafety* may lie in his personal encounters with disaster, his analysis points to the underlying threat of a more fundamental systemic breakdown arising from the unintended consequences of our contemporary designs for living. In holding the mirror up, he forces each of us to examine our responsibilities in these fundamental matters and challenges us to collectively scrutinise the relationship between science, society and humankind in our efforts at building a sustainable future. (Prof. Dr. Michael J Morley, University of Limerick, Ireland, President of IFSAM)

> In Unsafety, Professor Shigeo Atsuji has written a wide-ranging review of the dangers of inept, poorly-considered, profit-motivated managerial practices that have led to a series of modern 'disasters' in Japan and elsewhere in the world. While the origins of such disasters are often found in nature, it is the pathology of human organizations that is the focus of Atsuji's scathing criticisms. With particular emphasis on the recent nuclear disaster in Fukushima, he is careful to distinguish between the unavoidable dangers present in the natural world and the fully avoidable dangers that are inherent to a corporation mentality where individual managers, decision-makers and the executive elite are happy to share in the profits, but not in the burdens of systemic mistakes. A 'must read' for anyone interested in the future evolution of organizational management in a sustainable world. (Prof. Dr. Norman D. Cook, Kansai University, Japan)

> Prof. Atsuji is a unique scholar in social science. 'Unsafety' will make a significant breakthrough in normative sciences using interdisciplinary approaches. The authors of J. Reason's *Organizational Accident* and S. Bennett's *Disaster Management* are the pioneers of thinking on human accident and disaster. This book differs in presenting new issues: lost compliance in the Fukushima nuclear disaster, failure of management in the Japan Railways accident, the possibility of climate change from nuclear-heated oceans, and catastrophe arising from the linking of natural and unnatural disaster. (Prof. Dr. Koichiro Hioki: Management Philosophy, Kyoto University, Japan)

I wanted to make some return for the insights I have gained from overseas experts while presenting papers on the Fukushima nuclear disaster to the World Conference of the International Society for the Systems Sciences and the

International Federation of Scholarly Associations of Management. It is an honor for me to be published by Springer International Publishing for which I thank the Springer publishing editors Mr. Yutaka Hiraji and Ms. Shinko Mimura of TSS series. I would also like to recognize the contribution made to the publication process by Kazue Shinji and Thomas Hannon, who assisted with English language proofreading and translation, and by my doctorate students Kazunori Ueda and Ryōsuke Fujimoto, who provided collaboration. I appreciate their kind thoughts.

The present research is published in the *Translational Systems Sciences Series* by Prof. Hiroshi Deguchi (2004) and Prof. Kyouichi Kijima (2015), who are international systems scientists I respect. This book was supported by a scientific research grant from the Japanese government. Some of the book's research findings have been the subject of scientific papers. (MEXT KAKENHI Grant Number 24530437).

March 2015 Shigeo Atsuji

Contents

List of Figures

List of Tables

List of Photographs

Photographs by permission of:

List of Films

1 Masdar: Exploring Our Future, Saint Thomas Productions, France, 2012
2 The Future of Spaceship Earth, NHK, NEP/Primitive Entertainment, 2013–2014
3 Déchets: Le Cauchemar Du Nucléaire 25, Bonne Pioche/Arte France, France, 2009
4 What Happened at That Time? Series "Nuclear Power Plant Crisis, the Meltdown", NHK Special, 2012
5 Fukushima's Nuclear Disaster as Seen from the US., NHK, 2012
6 Inside Japan's Nuclear Meltdown, Quicksilver Media/WGBH, UK/USA, 2012
7 The Series "Meltdown File 2", NHK Special, 2012
8 Nuclear, Nothing to Report, Crescendo Films Iota Production, France/Belgium, 2009
9 Earth: Energy Quest USA, Passport to Knowledge Productions, USA, 2012
10 Can We Live Forever?, WGBH / Boston, USA, 2011
11 Why Poverty? Park Avenue: Money, Power & the American Dream, Democracy Pictures, NHK BBC DR ITVS SVT ZDF/Arte VPRO Steps International, 2012
12 Japanese Natural Resources at a Critical Juncture, Today's Close-Up, NHK, 2013
13 Mega-quake I, NHK Special, 2012
14 Overflowing Contaminated Water in the Fukushima Plant, Today's Close-Up, NHK, 2013
15 Can We Accuse the Company of Crimes? Eight Years After the JR Fukuchiyama Line Train Derailment Accident, Today's Close-Up, NHK, 2013
16 Series of Abnormal Weather Events: What Is Happening to the Earth?, August 29, 2013
17 Contaminated Water Crisis, Today's Close-Up, NHK, 2013
18 Contaminated Water: State of the Fukushima Plant, Today's Close-Up, NHK, 2013

19 3.11 Mega-quake, NHK, 2013
20 Loss of 320,000 Lives: How Can We Protect Lives from Mega-quake?, NHK Special, 2012
21 Terms and Conditions May Apply, Hyrax Films/Topiary Productions, USA, 2013
22 Google and the World Brain, Polar Star Films/BLTV, Spain/UK, 2012
23 The Sea of Japan: Gigantic Resources Unexploited in Its Deep Sea, NHK Special, 2013
24 Nuclear Terror: New Risk Threatening Japan, NHK Special, 2013
25 Continuous Typhoons: Why Such Unprecedented Torrential Rain?, NHK
26 Super-Typhoon in the Philippines, November 18, 2013, NHK
27 Report Update! Where Global Warming Goes, Science Zero, NHK, 2013
28 In Nuclear We Trust, Morgan Production/Kami Productions, France, 2013
29 The Atomic States of America, 9.14 Pictures, USA, 2012
30 Decommissioning Nuclear Power Plants: Mission Impossible?, Arte France/ Eclectic Presse, France, 2012
31 The Carbon Rush, Byron A. Martin Productions/Wide Open Exposure Productions, Canada, 2012
32 The Clean-Tech Future, VPRO, Holland, 2012
33 Power to the People, VPRO, Holland, 2012
34 Down-Winders: The Struggle of American Nuclear Test Victims, NHK Hiroshima, 2014
35 Why Do Renewable Energy Projects Not Advance, Today's Close-Up, NHK, 2014
36 How Will the 3.3 Trillion Yen Be Spent?: Monitoring the Recovery Programs, NHK Special, 2014
37 Series "Meltdown File 4: Vast Release of Radiation", NHK Special, 2014
38 Prevention Measures for Nuclear Accidents: Evacuation Plans, Today's Close-Up, NHK, 2014
39 Decisions of 130,000 Evacuees: Three Years After the Fukushima Accident, NHK Special, 2014
40 Mega-quake II-3: Prepare for the Worst Scenario, NHK and National Geographic Channels International (NGCI), 2012

About the Author

Shigeo Atsuji is professor of informatics at Kansai University in Japan and a research fellow at Kyoto University (2012–2013). With a DBA in organization theory and a Ph.D. in policy sciences, he is a member of a number of scholarly associations and societies. His research interests focus on management informatics and organizational intelligence, the decision-making theory, and the organizational aspects of accident and disaster, with an advisory board of KIRAS by FFG. Atsuji has presented associated case studies funded by the government at international conferences including those of the International Federation of Scholarly Associations of Management and the International Society for the Systems Sciences.

Part I
Disaster Chain

Contemporary society has viewed growth and prosperity alone as positives. Conversely, grieving and complaint have been seen as 'weakness' and 'loneliness'. The modern logic of economic growth has dismissed them as shameful and cast aside emotions essential to human nature. Actual human life is a mixture of contrasting emotions. It is by experiencing extremes of despair and passing through the depths of sorrow and suffering that humans reach the peaks of joy, and it is through the coming together of the opposites of joy and sadness that a profound humanity is gradually created. But in the materially oriented contemporary society where 'emotion' is a negative, true empathy and relationships of trust between humans are breaking down. In the market economy, the source of the problem can be traced to a social setup which fails to understand human emotionality. It is also reflected in the treatment of the individual which underlies contemporary society. This again raises the question of the conditions for human happiness, which can be seen to stand in opposition to human cooperation centered on organizations.

Chapter 1
Carbonized Terra: Paradox of Civilization

1.1 Genesis of Unsafety

Whether Asia or America, Europe or Africa, disasters happen all over the world and have been present at all periods in history, both recorded and unrecorded, and in societies of all creeds, whether Christian, Muslim, Buddhist, or of other religious identity. This book addresses the unnatural disasters and organizational accidents which arise as repercussions of natural hazards, revealing the current reality in Japan, where earthquakes, tsunamis, and typhoons are common, with particular reference to the Fukushima nuclear accident. Starting with the industrial revolution, modern social systems and administrative techniques have fostered the spread of unequal societies, a process which is continuing through the information and bioscience revolutions. Disasters are not merely an irregular factor affecting the economy, but more importantly cause great misery through loss of life, destruction of agricultural land, fisheries, and other sources of livelihood, and interruption of urban functions.

The unsafety caused by a disaster in one country also increases uncertainty in neighboring countries. As witnessed in the crop failures that preceded the French Revolution of 1789 related to the Mt. Laki eruption of 1783, African famines, and North Korea's drought, the food situation in a disaster-stricken country leads to famines, which can increase tensions with neighboring countries, and often precipitates wars or revolutions. If we go back further in history to the days of the nomadic Huns and Germanic tribes, what began with their raiding of neighboring grazing lands escalated into a great migration of peoples which threw the Eurasian continent into turmoil as described in *A World History* (W. H. McNeill 1999). A disaster can spark a chain of reactions that develops from physical damage through food shortages and refugee flows, international tensions, and increasing inequality, to economic depression and social unrest which spread to neighboring countries. In this way, a disaster can transmute progressively into a phenomenon with global repercussions. Historical evidence shows that disasters not only destroy safety but

S. Atsuji, *Unsafety*, Translational Systems Sciences 7,
DOI 10.1007/978-4-431-55924-5_1

have also been a harbinger of revolution and war. This aspect surely indicates the importance of disaster research.

Viewed from Japan at the far-eastern edge of Asia, Barack Obama, having ascended to the presidency some 150 years after the reconstruction of his nation under Abraham Lincoln, faces many problems. The gun culture and the money-worshipping market society that are spreading across America, the Land of the Free, like an infectious disease are exacerbating unsafety, damaging sustainability, and undermining the world's most affluent society. The prosperity created there by people from many different ethnic origins was akin to global biodiversity, and the creative ideas of its diverse population have made it a world leader. Its current direction is a terrible waste.

The German people made a rapid decision after the Fukushima disaster to abandon nuclear power, which is being implemented by Chancellor Merkel, while 90 % of the Italian people expressed the wish to end nuclear power in a referendum. These should be examples to Japan. Today, several years after the earthquake of March 2011, there are still 267,000 evacuees, of which 130,000 are people unable to return to their homes because of radioactivity from the Fukushima nuclear disaster. Meanwhile newspaper polls consistently show 75–80 % of the Japanese public to be in favor of ending nuclear power. Yet even a referendum is refused. The government, still controlled by industry and the businesses of major nuclear power nations, has adopted a policy of resuming nuclear power generation in the earthquake zone. The people are being made to suffer under the logic of industrial profit and commercial interest. Government decision-making without reference to the public has angered many people and caused exasperation. Japan now needs to invert its traditional wish to marry 'Japanese spirit with Western learning' by taking note of the wise decisions of countries which have maintained centuries-long prosperity.

In France, the world's second largest user of nuclear power after the United States, the pro-nuclear camp was defeated at the 2012 general election, with the defeat of Nicolas Sarkozy and election of President Hollande, and the position of the state-owned nuclear power company Areva was compromised. Iceland and the other Nordic countries are leading a shift to 'renewable energy'. A Swiss-type system of effective direct democracy, whereby the public are given a vote on individual policy proposals, is facilitated by modern information and communications technology, while the age-old representative system allows 'indirect democracy' to hold on. The safe and secure lifestyles of countries such as Switzerland, Austria, Bhutan, Canada, New Zealand, and Australia, which enjoy rich natural environments, are preserved thanks to the efforts of their peoples, who refuse to give in to the globalization of the market economy. In the future, what will be our needs, and what should be our demands? It is time to examine whether we really need the things prepared and provided for us. Emerging economies such as China, India, Russia, and Brazil—the so-called BRICs nations—are riding a wave of economic development, and their increasing electricity demand is driving a sharp increase in nuclear power station construction. However, a reading of history shows that the competition for energy and electricity resources has caused world wars and been a source of horror for humankind.

Far from the whole populace enjoying the benefits of such development, it only increases the fortunes of a small coterie of the rich, and wealth inequality widens further, while the gaps between north and south and east and west create imbalances and the imminent threat of unsafety. Japan, which experienced the great earthquake disaster of March 2011, is unparalleled not only in its marine resources such as fisheries, minerals, and energy but also in its highly refined technological abilities and human resources. Since the Fukushima nuclear disaster, even though almost all nuclear power operations have been halted (2011–2014), Japan's electricity supply has been maintained. Japan should now engage in initiatives of the kind possible only for a country that has experienced a massive earthquake of magnitude 9.0 and an accompanying tsunami. These should include the development of renewable energy, disaster- and crisis-management systems for urban infrastructure, technology and social systems for disaster prevention and mitigation, government policies to limit disaster and accidents, new and innovative systems for policy decision-making, crisis/risk-management approaches for business enterprises and local communities, and other sustainability-oriented measures which allow the world to benefit from the nation's tragic experience.

The present research casts a harsh light on the system error and human error that lie behind unsafety, the negative side of human cooperation that manifests itself in such forms as emotional confrontations, dishonesty, cover-ups, falsification, self-interest, and the shifting of responsibility. It represents a contribution from an interdisciplinary perspective to the idea that the accumulation of human-made disasters precipitates natural disasters. For this reason, it has some novel content, and although it represents the author's best efforts, there are also likely to be unintended mistakes and misinterpretations. On my journey backward through the natural and human-made disasters in the history of civilization, I will gladly accept correction from the reader. That the content of this book originates not in Europe or the United States, but in Japan, a country at the heart of the disaster zone, is to me a point of great significance.

1.2 Trans-field of Unsafety

In the wake of the unsafety episode represented by the Fukushima nuclear disaster, the aim of this book is to open an international debate, from a Japanese platform, on the interconnectedness of natural and human-made disasters. What is the ultimate origin of unsafety? From the standpoint of Japan, where the four tectonic plates around the Pacific Ocean meet and earthquakes, tsunamis, typhoons, and other disasters occur frequently, it is important to make the world aware of the correlation between natural disaster and human-made disaster. The unsafety in the title of this book applies not only to human error but also to system error at the level of social organization. Not only can natural disasters precipitate human-made disasters; the mechanism can also operate in reverse, with human-made disasters provoking natural disasters.

Historically, it has been known that natural disasters can lead to human-made disasters. According to W. H. McNeill's *A World History* [1], the great migrations of the Huns and other Germanic tribes were set off by droughts in their grazing lands which drove them into neighboring territory. Where natural disaster meant that hunter peoples, nomadic peoples, or agricultural peoples suffered drought in their grazing lands or crop failure leading to food crisis and were driven into neighboring lands, the result was sometimes war and conflict. The book's final Table 9.3 (Translational review between earth and human history) is a super-historical illustration of how natural disaster brings human-made disaster [2]. In the twice-repeated world wars, a quest for petroleum, coal, iron ore, and other energy or mineral resources led to conflict, a form of unsafety caused by humans themselves. Today, likewise, disputes over food resources, water resources, and energy and marine resources provoke needless international tensions in a cycle of sorrow and transience which appears to be in humanity's nature. A review even of European history shows that, starting with the French Revolution of 1789, related to poor harvests following the eruption of Mt. Laki in 1783, war, riots, internal strife, coups d'état, terrorism, uprisings, and other forms of disorder have almost always had their roots in crop failure and resulting food shortage, hunger, and famine. But in contemporary society, as illustrated in Chaps. 2, 3, 4, 5, and 6, the normal one-way flow from natural to human-made disaster has been supplemented by the paradoxical reverse of human-made disaster leading to natural disaster, creating another magnetic field of unsafety.

The alarm sounded over 'global warming', for instance in L.R. Brown's *Vital Signs: The Trends That Are Shaping Our Future* [3], indicates that, as foreseen in the sixth report to the Club of Rome, the environmental change brought about by human activity is beginning to have real repercussions. J. M. Diamond's *The World Until Yesterday* also suggests the undeniable possibility of a world that has become 'unsafe' [4]. The terra has alternated between 'snowball globe' periods, when the whole earth was frozen, and 'big melt' periods and is currently in an interglacial period. In addition to the resulting global warming and ocean warming, and apart from the natural disasters arising from atmospheric, marine, and geophysical changes or volcanic activity, there is also a spreading threat from the aging of vital infrastructure and 'systemic breakdown' exemplified by the weakening of social systems such as medical care, healthcare, pensions, welfare, and social insurance. Meanwhile, the population explosions in China, India, and Africa, which now account for half the world's total, will push the human population from 7.2 [5] to 10 billion (2050) according to the United Nations, provoking concerns over a potential food crisis.

Nowadays, the scale of human-made disasters is equal to that of natural disasters. Chemical pollutants such as PCB and dioxins, known collectively as 'environmental hormones', are on a sharp increase and have been reported as a cause of fetal brain damage, ADHD, and other physical and mental diseases. L. Brown's *Global Environment White Paper 2000* also warns that this negative legacy of the scientific revolution has contaminated the whole planet and is damaging humankind [6]. To pose the issue as characterized by R.B. Fuller, can Spaceship Earth—the global environment—support a population of over eight billion? The speed of climate change has outpaced predictions and is driving floods, drought, extreme cold spells, desertification, acid rain, and deforestation, exacerbating associated risk and crisis, and destroying the sustainability of humankind's existence.

The categories in the unsafety with which this book deals include not only natural disasters in the form of 'climate change' and abnormal weather patterns, earthquakes, tsunamis, floods, and drought but also human-made disasters such as those caused by corporate organizations' failed management with lost compliance and governance, military action, international tensions between states (governments) or other factors, and the arms trade and increasing militarization and military spending of societies in east Asia around China. Chlorofluoro-carbon gases may at first have been used only in small amounts, but through worldwide use they eventually created an unexpected hole in the ozone layer. This can be seen as a case of human cooperation accumulating to cause a natural disaster. Human cooperation brings economic growth, commercial profit, and social surplus, but on the minus side, it not only imposes burdens on the global environment [7], but is also a source of visible unsafety, for instance, through the imbalance of the growing wealth gap between north and south [8].

The trans-field of unsafety lies not only in physical aspects such as global warming, or chemical aspects such as environmental hormones, but also extends to biological aspects (viral infection, food crises, and competition for water and energy resources). As pointed out by J. M. Diamond in *Collapse* [9], the possibility of civilizational collapse precipitated by unnatural disaster cannot be ruled out given the examples of the Maya and Easter Island civilizations. In his *Guns, Germs and Steel*, [10] the unsafety whereby pathogens from a livestock and domestic-animal source infect humans with smallpox, measles, tuberculosis, whooping cough, influenza, malaria, or other diseases is presented as a case of a human-made disaster becoming a natural disaster. But more usually, natural and human-made disasters have led to catastrophe through a repeated process of action and reaction. In the face of the major floods that have occurred frequently across the globe in recent times, it might seem that the Old Testament tale of Noah's Ark is about to be realized. Fossil fuels such as petroleum and coal, together with forest resources which are subject to depletion due to their exploitation for timber and firewood, slash-and-burn agriculture, as well as dam construction, lead to the emissions of CO_2 and methane, resulting in Gaia's being carbonized. Sometimes in the near future, we will either have to start terra-reforming, or alternatively proceed to 'terra-forming' by moving to Mars.

In September 2013, the IPCC predicted that atmospheric temperatures will rise by 4.8 °C and sea levels by 82 cm by the year 2100. An environmental change unprecedented in human experience is in progress, and whether it consists of global warming in the current interglacial period, or of cooling, climate change is thought to cause crop failure and a resulting food crisis [11]. Here we see the connectedness of natural disasters with human-made disasters, based on a kind of 'negative cooperation' that produces CO_2, methane, and other greenhouse gases and PM2.5 in its wake. Today, global warming is manifesting itself in the extreme phenomenon of abnormal weather patterns worldwide, as witnessed in the anomalous cold spells and heavy snow that hit the eastern United States or the major floods that struck England in 2007 [12] and Wales in 2014 [13] and other events affecting the whole of Europe as well as Asia, Africa, and South America.

Below is a world hazard map showing the risk of natural and unnatural disasters. Dots on the map correspond to earthquake epicenters, tropical storms, drought, and the locations of the world's 441 nuclear power plants. Originally, the hazard map focused on natural disasters and was designed to indicate areas of risk and crisis. However, this fails to reflect war, ethnic conflict, religious tensions, global terrorism, and other anthropogenic factors of social, historical, or related origin. Today's worldwide hazards are not restricted to the perils of natural disaster. Rather, crises unleashed by unnatural disasters are of frequent occurrence and are expanding in scale at a global level (Fig. 1.1).

A look at the worldwide situation, with events such as the Gulf War and the Iraq War, the invasion of Afghanistan, the Arab Spring, the Syrian civil war, and terrorism outbreaks across the globe, shows a degree of complexity and confusion which can no longer be explained only in terms of the religious tensions between

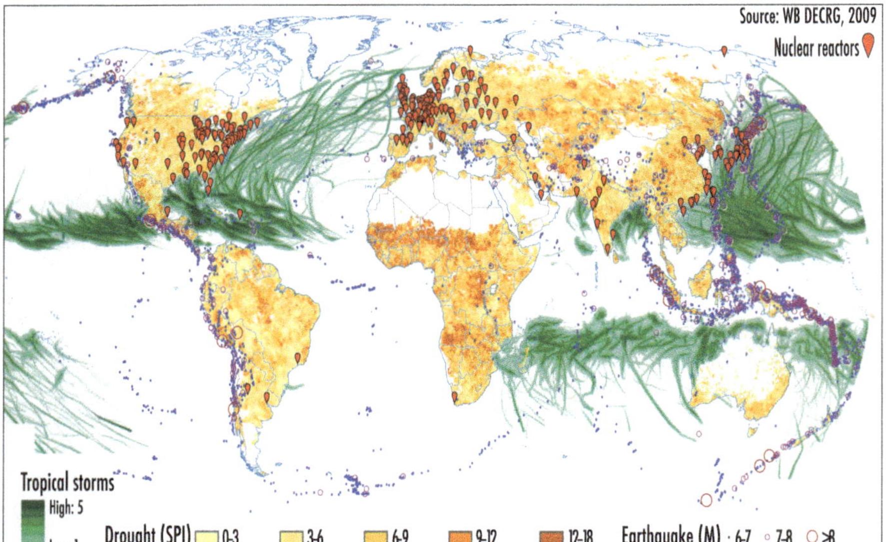

Fig. 1.1 Unsafety from natural disaster and nuclear power plants
Note: Mapping based on The World Bank, *Natural Disaster, Non-natural Disaster*, United Nations, 2011, p. 29
Location of world's 441 nuclear power facilities revised to reflect current situation
Adapted from Smolka, A.P., *World map of natural hazards,* Trans. Royal Society, A 2006; 364: 2147–2165

Christianity and Islam. In the Asian region, riots in various parts of China and the internal strife in Tibet, religious issues, and ethnic self-determination conflicts are compounded by the wealth gap. There is no end in sight to the expanding chain of resentment as people lose family and other loved ones to war and conflict. L.R. Brown's *Plan B 4.0: Mobilizing to Save Civilization* (2009) foresees the arms trade outstripping the grain trade, [14] with the escalating threat setting off a domino effect of increasing militarization. India's military spending has now reached a level equal to one-third of the world's total expenditure on the arms trade, with imports mainly from Russia leading to rising international tensions on the borders with China and Central Asia. The unnatural 'organizational accidents by human error' of which J. Reason writes conceal the danger of a disaster waiting to happen [15].

In 2013, news of the US National Security Agency's cyber-attack on Iran's nuclear facilities was published. On October 24 of the same year, Germany's Chancellor Merkel asked President Obama in the White House not to listen in on her mobile telephone calls. Inter-state 'cyber terror', especially by the USA and China, has escalated and increased global uncertainty. In addition to Iran's nuclear development, the 23 million people of a certain country with a standing army of 2.1 million are faced in a neighboring nation by an army of 1.2 million and 19 million reservists, making a total force of 20 million in a nation with universal conscription. In the seas of East Asia, there is increased international tension involving China, South Korea, Japan, Vietnam, and the Philippines regarding offshore rights to the Senkaku Islands, the island of Takeshima, and the area around the island of Hainan. On the other hand, America's call for an attack on Syria was rejected by the British parliament and Germany. This non-violent and non-militaristic approach has its historical parallels in Martin Luther King, Mahatma Gandhi, Nelson Mandela, and Aung San Suu Kyi. Today, we have perhaps reached the stage where a sustainable type of decision-making is required.

D. Meadows' *The Limits to Growth* (2004) suggests that the population explosion in Asia and Africa may make global renewal impossible [16]. In East Asia, international tension is being needlessly stoked over the economic rights to sea areas around remote and uninhabitable islands which do not even have a water supply, all because of the possibility of securing sometime in the distant future the rights to new energy resources like methane hydrate and mineral resources that will in any case be difficult to exploit. It might be expected that two world wars had taught us the foolishness of struggles over energy, mineral, and fishery resources, but such conflicts continue to be repeated. The warlike atmosphere could suddenly intensify. Interference and fabrication by third-party nations, seemingly intent on provoking war, are, as history shows, quite capable of resulting in a war situation.

The environmental destruction and damage to humans, not just from radioactive contamination from nuclear power stations but also from atomic bomb drops and hydrogen bomb tests, is incalculable and has also had a considerable impact on the global environment in the past. The Fukushima nuclear accident mentioned here has the dual aspect of a natural and a human-made disaster. Radioactive contamination equivalent to around 1,100 atomic bombs was concealed and falsified by central government and the electric power company, local communities suffered great damage, and the nation's entire population feared contamination. Additionally, there was the so-called 'damage by rumor' which attached to agricultural produce from the Fukushima area. This phenomenon represents a kind of unsafety. During the period of strong economic growth, when they established factories in China, Thailand, Vietnam, and several other nations, the Japanese became economic animals. Japanese society is like a school of fish *Gyogun* where everyone acts at the same time, all turning at once to the right when it is time to turn right and moving in unison to the left when it is time to turn left. This 'herd mentality' that make it impossible to turn back carries the seeds of risk and crisis. The Japanese conception of morality served the nation well in the postwar rebuilding and the recoveries after the Great Hanshin Earthquake of 1995 and the Great East Japan Earthquake of 2011, but in international relations the Japanese mentality has acted in unforeseen ways and produced unexpected situations, as history undeniably shows. Implicit in Japanese society is a fundamental ethos that rejects 'diversity'. In its handling of the Fukushima nuclear accident, the government was motivated by an attitude peculiar to Japanese culture, concentrating on mending international relations, but considering danger and failure as shameful and hiding them from local communities—its own people—who should have been informed.

1.3 Environment as Transitional Dimensions

What is the 'environment', which is the field of unsafety? The environment, sometimes in chaos and with a fractal aspect, cannot be understood solely in terms of a system. The environment including the ecosystem comprises the twin aspects of 'phase and corollary' and has a total presence, a total process, and a total structure. It is the field of dynamic phenomena that are spatial, historical, ecological, social, economic, political, and philosophical; it is both event and dimension: and, irrespective of micro- and macro-scopic distinctions, and without relation to organic or inorganic phenomena, it has relativity, continuity, and temporality.

According to the 'cosmic evolution hypothesis', [17], the multiverse that started with the Big Bang consisted of inflation and thermal energy only, but as the stars and planets materialized out of the gas, thermochemical reaction caused a progression from inorganic to organic which led to the birth of the phenomenon of life. The continued progression of the phenomenon of life led through homeostasis based on DNA copying to the advent of intelligence, which brought about the phenomenon of social organization. Human civilization itself is part of the continuum from matter through life to society, in which a simple world of thermal energy alone was transformed into the phenomenon of life in a process of progressively expanding 'complexity and diversity' that threw up the range of individual attributes (Fig. 1.2).

According to N. Wiener's *Cybernetics* [18], all phenomena share in the uniform diversification from matter through energy to information. Within the time-space field of cosmic evolution, unsafe history as shown in the hyper-chronograph views

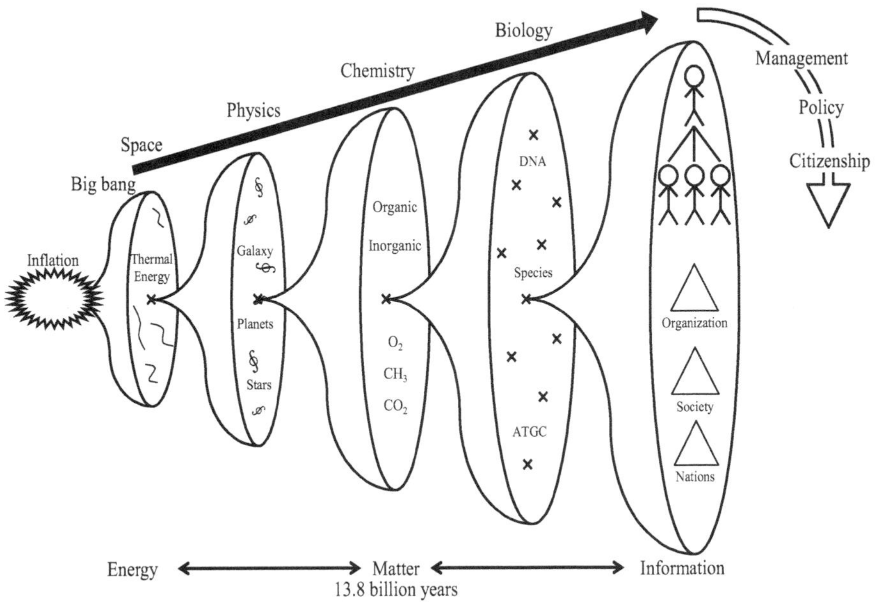

Fig. 1.2 Genesis by cosmic evolution: energy, matter, and information
Note: Inspired by "Complexity: cosmic evolution" by E. Chaisson, presented at ISSS Toronto 2000 [19]
See also, Table 1.1 Hyper-chronograph views of universe and earth

of the universe and earth (Table 1.1), unsafe history developed as shown including biological evolution led from single-cell organisms and other organic life-forms to the evolution of diverse plant and animal species, culminating in the development of neuron cells through which was acquired the ability to learn about and adapt to the external environment. Specifically, humankind created a social organism as a vehicle for cooperation, internalized the external environment as information, and also acquired a neural system for behavioral decision-making.

Biological evolution, seen as part of the cosmic evolution process, saw the transition from matter to biological organism and on to social organism, in which homeostasis has been maintained and complexity and diversity expanded [20]. As explained in items (1) to (5) below, the complexity of the world has evolved in stages from the physical to the biological to the social.

1. *Energy* ⇒ physical matter: the thermal energy of the Big Bang forced the expansion of space and, as the gas cooled, matter gave rise to the star and planet systems.
2. *Matter* ⇒ chemistry: thermal energy created space and matter, and chemical reactions among the stars and planets produced an increasingly diverse range of inorganic and organic substances.
3. *Chemistry* ⇒ genetic water-based life: the action of the diverse organic substances resulted in the creation of life, which acquired self-replicating ability

Table 1.1 Hyper-chronograph views of the universe and earth

Year (No. of years ago)	warm /cold	Event
13.8 billion		Birth of the universe
4.567 billion		Formation of the solar system
4.55 billion		Formation of the Earth
4.533 billion		Formation of the moon
4.404 billion		Oldest mineral deposit
4.1–3.8 billion		Late heavy meteorite bombardment
3.5 billion		Oldest life form
2.95–2.85 billion	cold	Pongola Ice Age
2.45–2.2 billion	cold	Huronian Ice Age Great Oxidation Event
2.02 billion		Vredefort crater
1.8–1.5 billion		Supercontinent Colombia
1.0–0.7 billion		Supercontinent Rodinia
730–635 million	cold	Snowball Earth
600–580 million		Emergence of multicellular life forms
580 million		Supercontinent Pannonia Gaskiers Ice Age
565 million		Ediacara biota
542–530 million		Cambrian Explosion
	warm	Emergence of fish
		Taconian orogeny
	cold	Late Ordovician Ice Age
444 million		**O–S extinction event**
	warm	Emergence of insects and land plants
		Caledonian orogeny
		Emergence of amphibians, fern forest
374 million		**F–F extinction event**
		Emergence of gymnosperms
		Acadian orogeny
	cold	Upsurge of Appalachian mountains (continental collision)
		Alleghanian orogeny
		Variscan orogeny
		Emergence of reptilians
	cold	Gondwana Ice Age
		Emergence of giant insects
		Formation of supercontinent Pangaea
251 million		**P–T extinction event (of the giant insects)**
199 million		**T–J extinction event**
		Cordilleran orogeny
		Expansion of gymnosperms
		Expansion of Navajo desert

■ global warming ■ global cooling

(continued)

Table 1.1 (continued)

Year (No. of years ago)	warm /cold	Event
		Archaeopteryx
		Giant dinosaurs
		Emergence of angiosperms
		Cretaceous period
66 million		**K–Pg extinction event (of the dinosaurs)**
		Diversification of mammals
		Cannonball Sea
56 million		Global warming maximum
50–30 million		Laramide orogeny
		Grande Coupure (great extinction)
		Emergence of great whales
		Widespread acid volcanic activity
		Ice Age
		Rapid upsurge of Himalayan mountain range
4 million		Mammoth
3.4–2.7 million		Australopithecus
2.1 million		Homo habilis
1.1–0.7 million		Java Man
790,000		Use of fire
600,000		Hewn stone tools
600–500,000		Peking Man
370–320,000		Nebraskan glacial stage
300,000–		Neanderthal man
280–230,000		Kansas glacial stage
195,000		Homo sapiens
180–127,000		Illinoian glacial stage
126,000		Sangamon interglacial stage (last interglacial stage)
90,000		Last glacial stage
74,000		Toba caldera
16,500	global cooling	Beginning of Jōmon Period
11,500	global cooling	Beginning of agriculture
8,200	global cooling	Destruction of Lake Agassiz
7,300	global warming	Akahoya eruption
5,500	global cooling	Sahara desertification

■ global warming ■ global cooling

Note: Based on "Complexity: cosmic evolution" by E. Chaisson, and *Super Time Scale of the Earth* by Japan Society of Geology [21]

through DNA, expanding the diversity of biological species. These primitive life-forms ingested external matter and acquired uncertainty.
4. *Life* ⇒ neural carbon-based species: in order for biological organisms to survive, they need to learn about the external environment, to which they cannot otherwise adapt. Life forms developed a neural system to internalize the external environment through symbols, acquired mechanisms for transmitting information, and began to select the items to be internalized.
5. *Neural system*: Information ⇒ thinking and intelligence: memory and learning are information functions serving the survival of the biological organism by symbolic internalizing against external phenomena. Survival ability has been enhanced by communicating knowledge learned from the environment and by cooperating. Meanwhile, life began to reshape the environment by not only adapting to the existing environment but also acting on and manipulating it. As an example, DNA bioengineering means that biological species can be changed as in L. Silver's *Remaking Eden* (1997) [22]. Already, a collective social intelligence has been acquired that works actively on the environment.

The ever-changing dynamic phenomenon of environmental diversity, from energy to matter to chemistry to life to neural systems to thinking and intelligence, can itself be seen as the holistic representation of the environment. From here, the human biological organism, by perceiving the environment and comparing and contrasting it with past memory, and by cooperating, creates social phenomena. The neural systems which learns about the environment, whether in a biological organism or a social organism, can be seen as the key element realizing homeostasis. Nowadays even the global environment is inconceivable without 'human will'. The cumulative effect of human cooperative behavior has reshaped the environment, but at the same time human-made disasters have emerged as a product of civilizational development. The cooperative behavior created by human collective consciousness does bring economic and social surplus, but can also transform into the cause of invisible disasters and accidents, the cumulative effect of which has, as history demonstrates, inflicted irreparable damage on the global environment. Because of the CO_2 emitted by burning timber and petroleum, coal, and other fossil fuels, 'carbonizing' the Terra of the globe has now begun.

The emergence of the world's complexity itself was not due simply to natural selection within the context of cosmic evolution but was assisted in a mutually complementary fashion by artificial selection through human forces based on the accumulative cooperative behaviors as posited by C. I. Barnard [23]. The concentration of human cooperation nowadays has the power to alter the global environment. The sum of human cooperation is reshaping the world artificially and transforming the global ecosystem. Humankind's advanced knowledge is no longer confined to the global ecosphere, but extends to space with the current exploration of the moon and Mars. There are 100 million items of space debris orbiting the earth at a speed of 7–14 km per second (far faster than the speed of a bullet), which, according to the 'Kessler syndrome' [24], have created a new potential unnatural disaster in the planet's orbit. We humans have expanded the reach of our 'visible hand' from the global environment to the planetary sphere. As a result, the negative aspects of human cooperation need to be 'X-rayed' for unsafety. We are used to seeing our state of health through the indices of economic growth, GDP and GNP (visible light), and entropy, but if instead we were to survey the world through the X-ray mode of societal unsafety, or enthalpy, what would show up? A function for 'revealing the invisible' by identifying unsafety through the implicit signs latent in disaster and accident situations may be necessary for a sustainable society going forward.

As at the time of the great centuries-long migrations of the Huns and other Germanic tribes, today the global population is in a state of explosion, precipitating a food crisis. The drying-up of water resources and the competition for energy resources is intensifying and China's 1.6 billion people and India's 1.3 billion are in overspill, pushing west, east, and south and setting off another great movement of peoples, which is sending a wave of unsafety not only through neighboring nations but across international society as a whole.

References

1. McNeill, W.H., *A World History*, Oxford University Press, 1998.
2. Bennett, S., *Innovative Thinking in Risk, Crisis, and Disaster Management*, Gower Publishing, 2012.
3. Brown, L.R., Flavin, C. and Kane, H., *Vital Signs: The Trends that Are Shaping Our Future*, Worldwatch Institute, 1992.
4. Diamond, J.M., *The World Until Yesterday: What Can We Learn from Traditional Societies?* Viking Penguin, 2012.
5. Smith, L.C., *The World in 2050: Four Forces Shaping Civilization's Northern Future*, 2010.
6. Brown, L.R. et al., *State of the World 2000*, W.W. Norton, 2000.
7. Meadows, D.H., Meadows, L.M., and Randers, J., *Beyond the Limits*, Chelsea Green, 1992.
8. Sachs, J.D., *The End of Poverty: How We Can Make It Happen in Our Lifetime*, The Wylie Agency, 2005.
9. Diamond, J.M., *Collapse: How Societies Choose to Fail or Succeed*, Viking Penguin, 2005.
10. Diamond, J.M., *Guns, Germs, and Steel: The Fates of Human Societies*, W. W. Norton, 1997.
11. Cribb, J., *The Coming Famine: The Global Food Crisis and What We Can Do to Avoid It,* University of California Press, 2010.
12. Wilby J., Neal, R. and Bell, S., "Emergent Disaster Response during the June 2007 Floods in Kingston upon Hull, UK", *Journal of Flood Risk Management*, Wiley, 30 August Online, 2011.
13. Tunstall, S.M., Johnson, C. and Penning-Rowsell, E. *Flood Hazard Management in England and Wales: From Land Drainage to Flood Risk Management.* World Congress on Natural Disaster Mitigation, 19–21 February 2004, New Delhi, India.
14. Brown, L.R., *Plan B 4.0: Mobilizing to Save Civilization*, W.W. Norton & Company, 2009.
15. Reason, J., *Managing the Risks of Organizational Accidents*, Ashgate Publishing, 1977. *See also*, *Human Error*, New York: Cambridge University Press, 1990.
16. Meadows, D.H., Randers, J. and Meadows, D.L, *Limits to Growth: The 30-Year Update*, Chelsea Green Publishing, 2004.
17. Chaisson, E., "Order, Flow and Complexity: A Precis on Cosmic Evolution", *Proceedings of World Congress of System Sciences 2000*, ISSS, pp. 22–23.

18. Wiener, N., *Cybernetics or Control and Communication in the Animal and the Machine*, MIT, 1948.
19. Chaisson, E., "Complexity: cosmic evolution", *the 45thISSS Proceedings*, International Society for the Systems Sciences, 2000.
20. Krauss, L.M., *A Universe from Nothing: Why There is Something Rather than Nothing*, Simon & Schuster Ltd, 2012.
21. Japan Society of Geology (supervised), *Super Time Scale of the Earth*, Iwanami, 2014.
22. Silver, L.M., *Remaking Eden: How Genetic Engineering and Cloning will Transform the American Family*, Avon Books, 1997.
23. Barnard, C.I., *The Functions of the Executive*, Harvard University Press, 1938.
24. Kessler, D. J. and Cour-Palais, B.G., "Collision Frequency of Artificial Satellites: The Creation of a Debris Belt", *Journal of Geophysical Research* 83, 1978.

Chapter 2
The Fukushima Nuclear Catastrophe: Systemic Breakdown and Pathology

The magnitude 9.0 earthquake and tsunami that struck northeast Japan on March 11, 2011, were unavoidable natural disasters, but we consider the subsequent breakdown of the Fukushima nuclear reactors to be a catastrophe created not only by nuclear-engineering systems but also by avoidable system errors. This chapter [1] provides an analysis of the organizational disaster which lies behind the 'systems pathology' of problems at the Fukushima Daiichi (number one) Nuclear Power Plant (hereafter 'Fukushima Daiichi Plant') of Tokyo Electric Power Company (TEPCO): the operation of old reactors, the concealment of incident data, and the problems of supervision including the 'nonrational' location of the power station. Secondly, we consider the mechanism of 'catastrophe' [2] based on the Fukushima case. Finally, I suggest a rational proposal concerning the phaseout of nuclear power around the world. This proposal is made with a view to obtaining sustainable decision-making for our future, not simply in light of the supply-and-demand decisions around electrical power, but also in consideration of the homeostasis of environmental aspects including ecosystem-based social systems, and their contribution toward our sustainability. It should be noted that this chapter is not intended as a particular criticism of the electric power companies and Japanese government but rather an analysis of what went wrong in the Fukushima Daiichi Plant accident.

Reworking of: Atsuji, S. et al., "The Fukushima Catastrophe seen as a Malfunction of Organizational Systems: Nuclear Fade-out and Hazard Maps for the World", *General Systems Bulletin*, Vol. XXXXI, 2012, pp. 26–30.

S. Atsuji, *Unsafety*, Translational Systems Sciences 7,
DOI 10.1007/978-4-431-55924-5_2

2.1 'Threefold Disaster': Earthquake → Accident → Human-Made Disaster

Reflecting on the disastrous earthquake and nuclear accident of March 2011 in Japan (Photograph 2.1), author Hiroyuki Itsuki invokes the 8th-century Chinese poet Toho with his paraphrase of the poet's famous words: "Though its mountains and rivers be destroyed, yet the country remains." He casts this as a 'second defeat' comparable to that of World War II—but at that time, in contrast, it echoed Toho's original words: "The country destroyed, yet its mountains and rivers remain." The gigantic earthquake and its tsunami engulfed the material, biological, and social systems that people had built up over the course of their history. Today what remains is a cultural system of human interrelation, which has become the driving force behind restoration and reconstruction. The disaster at TEPCO's Fukushima Daiichi Plant brought about a Promethean-like catastrophe from the 'fire' of the atom, the putatively 'unexpected results' of technological systems. Despite the power company's having in place a social system comprising organizational systems and administrative supervision, which should have controlled the atomic flame, the scope of the accident widened, like that of the Chernobyl [3] and Three Mile Island accidents.

While large-scale earthquakes and tsunamis are natural disasters that are difficult to evade, the Diet (Japan's houses of parliament), private accident-investigation commissions, and even the Nuclear Safety Commission all concluded that the

Photograph 2.1 Miyako tsunami *Source:* by *permission* of Miyako city office authority

nuclear accident was partly caused by human's system error, and that there was the possibility of having avoided it. This chapter examines such issues as the government response to the Fukushima nuclear accident and the background of the power company involved, viewing this catastrophe from the viewpoint of human-made disaster as 'an aspect of systems pathology' [4], with the aim of preventing such large-scale tragedies in the future. Herein we shall elucidate the mechanism leading to the worsening of the accident to the level of a catastrophe, formulate a disaster index related to the nuclear accident, and study the upcoming 'Draft Plan of Power Scenarios' from the aspect of public welfare.

The Fukushima nuclear accident was initially precipitated by the Great East Japan Earthquake and tsunami, both disasters then proceeding simultaneously with a third, human-made one as the government and power companies directed inhabitants to evacuate as part of their accident response. The disaster was threefold (Fig. 2.1): the earthquake and tsunami were the first disaster, the nuclear accident second, and then the evacuation instructions and other responses to the accident the third propagation of the disaster. At first glance, the earthquake and nuclear accident can be regarded as one entity; however, discrete event analysis reveals that what may appear to be instantaneous events actually operate both at a level of interrelated, hierarchical phenomena and a level at which independent phenomena are simultaneously linked. If one traces back the chain of disasters to the background which elicited them, a pattern of disaster becomes visible, incorporating such components as the aging of the Fukushima Daiichi Plant's six reactors over 40 years of operation, a longstanding history of trouble, the concealment of such incidents and/or falsification of attendant data, right back to the historical circumstances surrounding the siting of nuclear plants in an earthquake-prone region and concomitant with these, defects in the crisis management and response of the authorities. Now the systems pathology concealed in the technological, organizational, and social systems that are the basis of the nuclear power industry has come to light in the 'catastrophization' arising from the chain of accidents.

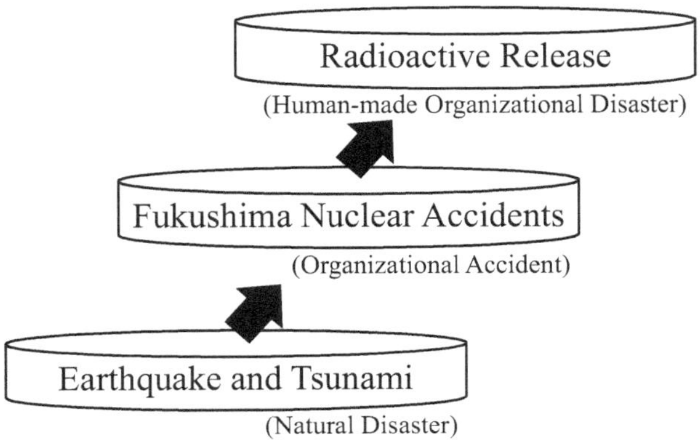

Fig. 2.1 Threefold disaster of the Fukushima Daiichi Plant accident

Sitting on the four tectonic plates of Asia, Eurasia, North America, and the Philippines, Japan is a disaster-prone country susceptible to intraplate earthquakes, tsunamis, and typhoons. In geopolitical terms, it is surrounded by powerful nations, with Russia to the north, China to the west, and North America across the sea to the east. It is therefore in a location where it pays to be careful.

Oddly perhaps, it was this country, already witness to the Hiroshima and Nagasaki atomic bombings, which was struck by a globally unprecedented nuclear accident—namely, the Fukushima disaster—triggered by a major earthquake and tsunami. It was subsequently revealed that the Fukushima nuclear power station had been at the center of repeated cover-ups, misinformation, and deceptions and that these and other organizational failings and the Japanese government's ill-defined regulatory system for nuclear power had been compounded by the unfortunate siting of a nuclear facility in an earthquake-prone region. Tracing the origins further back, there were found to have been secret agreements with the United States on nuclear strategy and a number of facts kept hidden from the Japanese public.

Seen historically, the Fukushima nuclear disaster was not simply a natural disaster but was also meshed together with a human-made disaster caused by a cumulative series of accidents. It was in fact a threefold disaster combining aspects of nuclear technology, organizational control, and energy policy, each inherent in the nuclear power system. It was a catastrophe waiting to happen. This unprecedented disaster has exposed the unsafety of the nuclear power system not only in Japan but in all countries operating nuclear power stations. As for nuclear power as an energy source to meet the electricity demands of the developed nations, the question of its vulnerability has been raised in terms of the safety of its entire hierarchy of systems, including technological systems, safety-control systems, and nuclear power regulatory systems.

2.2 The Background to the 'Accident Catastrophization' Visible in the Fukushima Daiichi Plant

2.2.1 Irrational Siting for Nuclear Power: Operating in an Earthquake-Prone Region

Japan's topography places it at the junction of the Eurasian, North American, Pacific, and Philippine tectonic plates (Fig. 2.2): a site for numerous earthquakes. The Pacific side of the recently seismically active Tōhoku region had often in the past been struck by large-scale earthquakes of magnitude 6 or greater. Further, strata analysis has established that the Jōgan Earthquake, one of the largest on record, took place in the region in AD 869, and the area where the Fukushima plants were constructed had a high risk of both earthquakes and tsunamis (Photograph 2.2) on the Pacific side. Seismic data from 2010 recorded 226 earthquakes in Miyagi Prefecture and 128 in Fukushima Prefecture, of which four instances in both

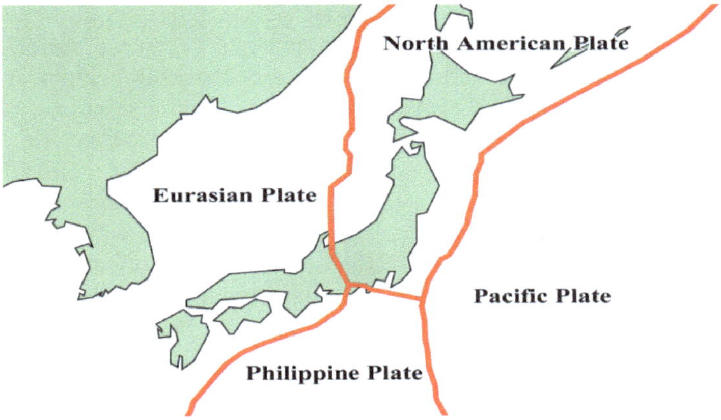

Fig. 2.2 Four plates: Eurasian, North American, Philippine, and Pacific

Photograph 2.2 After tsunami 2011 *Source*: by *permission* of Otsuchi town office authority

prefectures were '4' or higher on the Japanese seismic scale of 1–7 (7 being the highest) [5].

Why did the government give permission to site and operate an atomic power plant in an earthquake-prone region? In retrospect, it appears that to cover the demand for electricity in metropolitan areas, commercial and economic factors were prioritized over safety in the political decision-making process determining the siting of nuclear power facilities. For example, it became a societal given through the subsidy system under the Three Laws for Power Development for the many fiscally straitened local municipalities to accept the siting of nuclear plants in exchange for compensation for local citizens underwritten by the state. This was by no means limited to the Fukushima and Ikata nuclear plants. This political prioritization of power provision in urban areas can be identified as a background factor in the trend toward the disregarding of the problems associated with siting nuclear plants in earthquake-prone regions. It is an undeniable historical situation that during the period of high economic growth in Japan in the 1960s to '70s, with its attendant demand for power, not only power companies but state policies prioritized the maintenance of an electricity supply based on nuclear power.

2.2.2 The Limits to Control Apparent in Nuclear Power System Error

Japan has 54 nuclear reactors, of which one-fourth was constructed in the 1960s or '70s, and many are thus aging facilities that have been in service for 30 years or more. The service life of the main component of the nuclear reactor, the pressurized container vessel, has been established in Japan as 40 years, limiting operations to this length. However, in February 2011, 1 month prior to the accident, the Nuclear and Industrial Safety Agency (NISA) decided to "extend operational life from 40 to 60 years" [6]. Further, in October 2010 the Japan Nuclear Energy Safety Organization (JNES) reported on the "danger of shell melt-through 1 h and 40 min after loss of power supply" [7]. Compared with a meltdown, in which the reactor core melts within the pressurized containment vessel, a melt-through, in which the core breaches the exterior of the containment vessel, is far more dangerous. Further evidence that such problems were receiving increasing attention can be found in the Citizen's Nuclear Information Center report "Can the Fukushima Plants withstand an Earthquake?" [8].

Table 2.1 displays the years of operation of the aging nuclear plants and the number of incidents, including those that may have comprised a radiation leak. Particularly salient are the total 120 instances of failure at the Fukushima Daiichi Plant prior to the March 11 disaster. One aspect of aging reactors is the varying characteristics of the different kinds of metals and other materials employed in their construction. With the passing years, factors like corrosion and structural fatigue undermine the durability of such materials, with the attendant potential for malfunctions. This closely parallels the pathologies that manifest as a human ages. The Fukushima Daiichi Plant's reactors often experienced problems with system safety management such as 'cracks in the reactor containment wall' and 'loosened bolts', resulting in levels of danger that set off alarms prior to the disastrous earthquake [9].

Table 2.1 Age-related deterioration of reactors and number of instances of failure (March 2011)

Plant Name	Period of Operation	No. of Incidents
Tsuruga Plant No. 1 Reactor	42 years	81
Mihama Plant No. 1	40 years 4 months	32
Fukushima Daiichi Plant No. 1	40 years	54
Mihama Plant No. 2	38 years 8 months	37
Shimane Plant No. 1	37 years	17
Fukushima Daiichi Plant No. 2	36 years 8 months	57
Takahama Plant No. 1	36 years 4 months	46
Genkai Plant No. 1	35 years 5 months	24
Takahama Plant No. 2	35 years 4 months	41
Fukushima Daiichi Plant No. 3	35 years	31
Mihama Plant No. 3	34 years 3 months	38
Ikata Plant No. 1	33 years 6 months	27
Fukushima Daiichi Plant No. 5	32 years 11 months	21
Fukushima Daiichi Plant No. 4	32 years 5 months	20
Fukushima Daiichi Plant No. 6	31 years 5 months	29

Source data: Y. Masai [5]

All six of the Fukushima Daiichi Plant's reactors commenced operation more than 30 years ago, and their operations were extended. One of the types of reactor employed is the GE Mark I, with a containment vessel smaller than that involved in the Three Mile Island nuclear accident. US nuclear technicians and researchers have pointed out defects in its structure. One may posit that the combination of such structural defects and aging characteristics meant that the reactors were unable to withstand the unexpected effects of the earthquake and tsunami, thereby exposing 'administrative limits'.

2.2.3 Deterioration of Nuclear-Management Systems: Cover-Ups and Falsifications

The safety management of nuclear reactor systems does not consist of technological and engineering aspects alone. Despite the frequent instances of failure with the Fukushima Daiichi Plant's six aging reactors, it is known that cases that should have been reported to the regulatory authorities were concealed, and that organization-wide falsification of data occurred. Table 2.2 (inappropriate handling of incidents at the Fukushima Daiichi Plant) demonstrates the avoidance of not only the safety management of nuclear operation but also of public disclosure that should have been made to plant workers as well as local residents. For example, cracks in the 'shroud' (reactor containment wall) were altered in accident data to appear as

Table 2.2 Inappropriate handling at the Fukushima Daiichi Plant

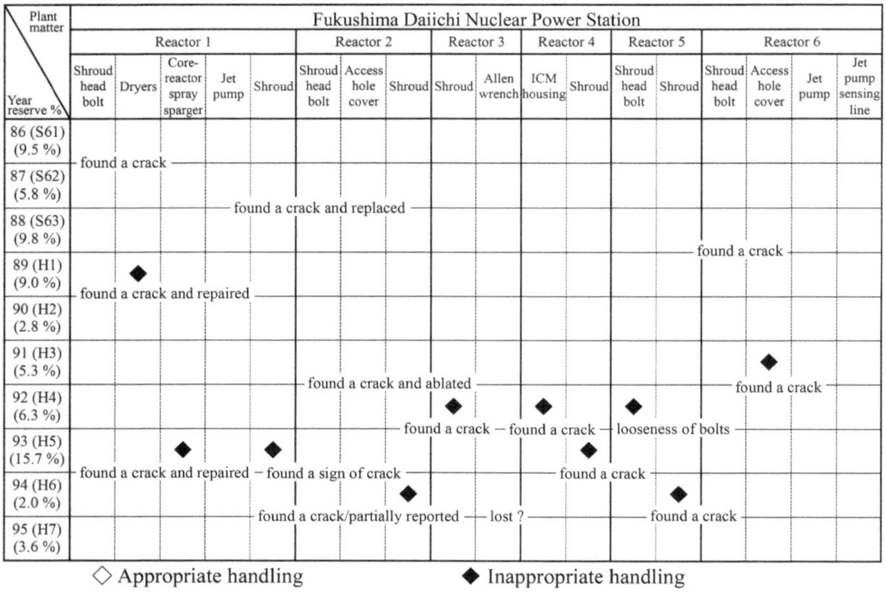

Source: Nihon Kogyo Shimbun [9]

'loosening in bolts' [10]. The state of organizational management, with its downplaying of serious trouble, fabrication of data, and inadequate reporting, calls into question not only the electric power industry's business ethics but also the industry's corporate governance with regard to nuclear power.

A fundamental issue of human resource management in the realm of energy has concerned each power company's safety management of manual workers at their nuclear plants. One cannot ignore the social responsibility the power industry as a whole bears in their condoning of the reassignment to different electric-power jurisdictions of nuclear-plant employees whose radiation doses exceeded set limits. The contradiction inherent in nuclear power's industrial application consists in how the 'limits of organization' ensnare all stakeholders: not only manual workers at atomic plants, subcontracting businesses, and local residents but also divisions involved in safety control, the management ranks, and administrative authorities struggling to provide explanations to the nation.

The safety of atomic energy was previously brought into question by the JCO criticality accident in the town of Tokai, Ibaraki Prefecture, in 1999 (Chap. 6). Essentially, since the JCO facility was not a nuclear power plant, the accident simply concerned the manufacturing of fuel for nuclear power, and evidently the term *sōteigai* 'beyond expectations' was used. As a result, a review of the overall administration of nuclear power was conducted, which putatively led to the transfer of control to the present organization. However, the lack of transparency following the 3.11 disaster in the sequence of actions and disclosures of information of the bodies charged with nuclear safety—namely, the Ministry of Economy, Trade and Industry (METI) and Science and Technology Agency running the Nuclear and Industrial Safety Agency (NISA), and the Nuclear Safety Commission—demonstrated a situation similar to that of the 1999 JCO criticality accident. What these two incidents have in common is the 'negative structural inertia' that can be identified as pervading the interdependent social system of industry, government, and academia: a phenomenon that inheres in the ingrained habits of the power industry and nuclear power administration.

With regard to this structural inertia, the present author S. Atsuji together with former power-company technician S. Senba outlined in the 2002 Association for the Study of Industrial Management (Japan) annual review, *Liberalization of the Power Industry and the Public Sphere*, the circumstances under which the power companies repeatedly invested in generation facilities despite an oversupply of electricity [11]. On display was the power industry's loss of a sense of 'social well-being', with the nuclear power industry's monopoly and the entrenched vested interests that characterized the power industry in general exacerbating this negative structural inertia.

2.3 The Mechanism of the Fukushima Daiichi Plant Catastrophe

2.3.1 *Systems Pathology in Organizational Disaster*

The Diet Accident Investigation Commission's published report detailed the Great East Japan Earthquake and subsequent Fukushima Daiichi Plant accident (Photograph 2.3) [12], as well as the deficiencies in the government's and power companies' responses in terms of evacuation instructions. When asked by media outlets whether the event was an accident or a disaster, Nuclear Safety Commission chairman H. Madarame declared it a 'human-made disaster'. The Fukushima Daiichi Plant disaster, initiated by the earthquake, can be regarded as an event in which the natural disaster and nuclear accident mutually aggravated each other. The human error manifested at each system level, and the system error latent within it, compounded each other. The security holes that caused the errors were not limited to safety management, extending as they did through every system layer up to organizational management and policy decisions. In summary, the 'spiral of failure' [13] comprised the following elements: first, the irrational siting of a nuclear power facility in an earthquake-prone region (a policy system issue); second, continued approval of the operation of aging reactors; third, power companies' concealment of instances of failure and falsification of related records (degradation of organizational management systems), and in addition structural inertia observable in electric power policy and nuclear power administration (social-systems pathology).

Photograph 2.3 Fukushima Daiichi Plant *Source:* by *permission* of Tokyo Electric Power Company authority

J. Reason in his *Organizational Accidents* compares an organization to processed cheese: the causes of an accident infiltrate the 'security holes' in the organization, leading to a disaster [14]. Evidently one characteristic of the Fukushima Daiichi Plant accident is that it must have circumvented the technological, organizational and social protection systems, leading to the threefold disaster. The baleful synergy of the accident and human-made disaster arose from the coupling of the effects of the huge earthquake and tsunami with malfunctions in the aging nuclear system, and was a consequence of their penetration of the security holes in the presiding administrative body. Overall the scale of the disaster enlarged, with the natural and unnatural elements operating together to inflict comprehensive damage encompassing the physical, geographic, biological, and social systems. Presently the sole-remaining intact system is the cultural system of interpersonal 'bonds', wherein human interdependence and trust are the key to restoration and reconstruction.

In order for the next generation to learn the lessons from the Fukushima Daiichi Plant disaster, its elements should be conceptualized along the lines of the concepts in Reason's 'Unsafe Acts' [15] by integrating the criteria of interdependent 'physical, biological, and social limitations' [16] so as to depict the 'catastrophe mechanism' as in Fig. 2.3. This figure illustrates the characteristic pathology of the processes that play out in a catastrophe, whereby each system layer is paralyzed, and the system itself collapses into sudden dysfunction or a state of panic. In the figure we see the dynamism of a 'chain of disasters'—moving from natural disaster

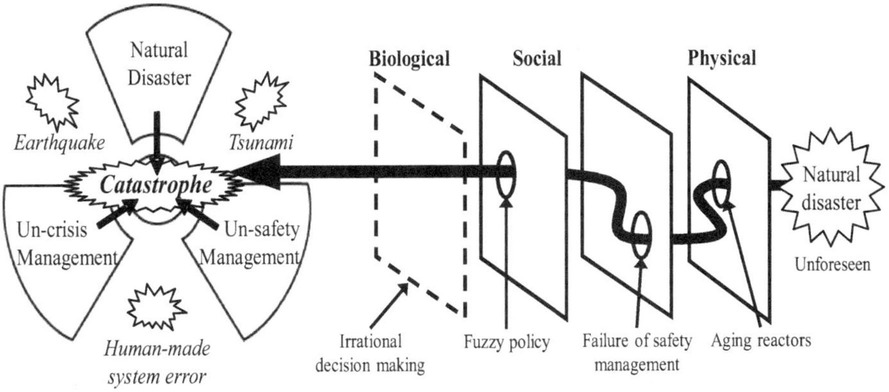

Fig. 2.3 Catastrophe mechanism *Note:* Adapted from J. Reason's 'Swiss Cheese Model' of the Organizational Accident occurring through the 'Security Holes'

to the cause of the accident, then finally into human-made disaster—which circumvents the technological and social protection systems and becomes a unified whole.

2.3.2 International Comparison of Nuclear Accidents/ Disasters

In analyzing the Fukushima nuclear disaster, we have compared it with the international precedents (Table 2.3), which are clearly (1) the accident at the Three Mile Island nuclear plant situated on a fluvial sandbar between New York and Washington in 1979 [17], when a reactor melted down and leaked radioactive substances into the environs; and (2) the Chernobyl nuclear accident in the former Soviet Union in 1986 [18], which disseminated large quantities of radioactivity.

Common to the Three Mile Island, Chernobyl, and Fukushima accidents are mistakes made by personnel. This human error caused system breakdowns at each level, and crisis management was similarly unresponsive. Another commonality is the lack of public information provided about the accidents, not only to residents living in accident areas but to the public as a whole: from Three Mile Island, to Chernobyl, and now onto Fukushima, scant lessons have been learned. One can identify in these three nuclear disasters a pathological syndrome in which human error—as in the operational errors at Chernobyl and the gauge misreading at Three Mile Island—precipitated a panic situation, further exacerbating the organizational accident.

However, Fukushima Daiichi Plant accident data comes from the NISA report entitled "Tōkyōdenryoku kabushikigaisha Fukushima Daiichi genshiryokuhatsudensho no jiko ni kakawaru 1gōki, 2gōki oyobi 3gōki no roshin no jōtai ni kan suru hyōka ni tsuite" [Assessment of the Condition of Nos. 1, 2 and 3 Reactors Affected in the Tokyo

Table 2.3 International comparisons of major nuclear accidents

	Fukushima	Chernobyl	Three Mile Island
Date	March 11, 2011	April 26, 1986	March 28, 1979
Accident details	Damage to reactors 1–4, loss of cooling systems, core meltdown	Reactor four runs out of control during test operation and explodes	Mechanical failures lead to cooling system failure and core meltdown
Amount of radiation released	11,347,000 TBq	13,194,000 TBq	9.8 TBq
INES rating	7	7	5
Region affected	Entry prohibited within a 20-km radius	Compulsory evacuation within a 30-km radius	Evacuation advisory within an 8-km radius
No. of evacuees	Estimated approx. 140,000	Approx. 135,000	Approx. 144,000
Accident causes	Damage to reactors and cooling systems due to earthquake and tsunami; subsequent delay in operation of emergency cooling system	Runaway reactor owing to compounded design errors, defects in the manual, and on-site command errors (Inappropriate response/operational errors)	Cooling system failure due to mechanical breakdown, gauge misreading and operational errors (Gauge misreading/operational errors)

Source: United Nations Scientific Committee on the Effects of Atomic Radiation, Sources and Effects of Ionizing Radiation, Report to the General Assembly with Scientific Annexes, UNSCEAR 2008 Report vol. II, 2011, pp. 70–71

Electric Power Company's Fukushima Daiichi Nuclear Power Plant Accident], October 20, 2011.

What must not be forgotten is the reality that the Fukushima Daiichi Plant accident constituted an unprecedented complex of simultaneous disasters: the six reactors came to a halt, three of them were wracked by intermittent explosions and went into meltdown, and there was a continuous release of radiation, expanding the irradiated area. There are many historical cases of disaster in which the vulnerabilities and security holes of safety systems, organizational management, and social systems compound each other and result in the unanticipated catastrophization of an accident.

2.3.3 *Measurement of System Degradation: Formulation of the Disaster (and Reactor Decommissioning)*

As previously mentioned, we examined the mechanism leading to the Fukushima Daiichi Plant accident by following the chronology from the background to the accident on to the accident itself. Further, post-accident response is of special note in terms of the inadequacy of evacuation directives and advice regarding the direction in which the radiation was spreading. The Japanese and German meteorological agencies gave completely opposite advisories for the direction in which the wind was blowing at the time of the accident. As a result, the radiation contamination estimates from the German meteorological agency, which had experienced the effects of radiation from the Chernobyl accident, accorded with the more than 60 items of SPEEDI data that Japan possessed but did not release at the time. It is no coincidence that Japan, the only country to have undergone atomic bombing, has suffered the fate of nuclear disaster both at Hiroshima and then at Fukushima.

An overview of the world's 441 nuclear reactors at the time of writing indicates that there are no standards for the decommissioning of aging reactors: on the contrary, the subject is regarded as taboo. The US Atomic Energy Commission, while not talking of Japan's 'life extension to 60 years', has in fact notified the 'possibility of a 20-year extension', meaning that the length of extension for continuous operation is effectively the same (*The Wall Street Journal*, July 21, 2011). The rise in the number of cancer patients is proportionate to the nuclear tests conducted both by the United States and Soviet Union and the development of atomic facilities that basically began with the Japan–United States Atomic Energy Agreement. Analogous is the rise in endocrine-system-related illnesses proportionate to the accumulation in the environment of hormone-disrupting chemicals, identified by T. Colborn, which notably displays the same incremental curve when graphed.

As an extreme generalization, an arguable paradox exists that, as with the nuclear accidents in the US and former Soviet Union, historically nation, states and organizational systems do not protect individuals. Such systems are designed to be conducive to people's welfare, but characteristically the processes the system operates as a whole are geared toward self-preservation, with the result that individuals and the whole are alienated from each other. This applies equally to nuclear power systems. With the complete lack of standards for reactor decommissioning, nuclear power plants that have been operating for 30 years or more are indubitably the very manifestation of this system paradox. In light of this present situation, it seems that a scale or index that can inform residents of the level of risk is essential for human homeostasis.

The antithesis inherent in the nuclear power system—that is, economic factors versus risk factors, and convenience versus fear—is actually two sides of the same coin. For that reason, in order to quantify the elements of risk and fear latent in such factors as economy and convenience, we wish to attempt to formulate a disaster index based on the data released about the Fukushima Daiichi Plant before the accident that residents can use to calculate risk.

If we metaphorically relate this nuclear power disaster formula to the characteristics of human illness, then, in the formula shown as *Formula 2.1*, time-related degradation of nuclear reactors is α ('aging'), frequency of past instances of failure is β ('history of illness'), precipitating factors like earthquakes are γ (the 'trigger'), safety-management systems are ρ ('chronic care'), and the credibility of administrative supervision is τ ('care policy'). One does not need to be an expert to understand this formula: residents

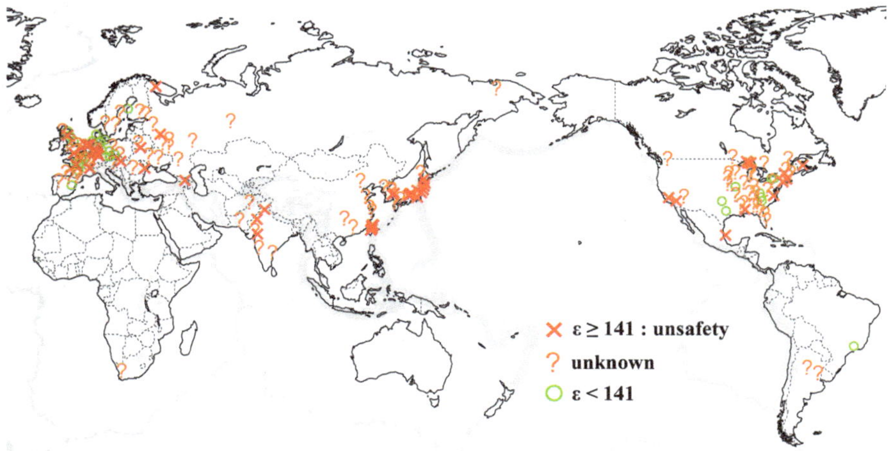

Fig. 2.4 World nuclear hazard map *Note:* Mapping based on the World Nuclear Database (URL: http://world-nuclear.org/NuclearDatabase/) and the Japan Nuclear Energy Safety Organization: "Kokunai/kokugai toraburu jōhō" [Information on Domestic and International Instances of Failure] (http://atomdb.jnes.go.jp/)

of a region, including children, can calculate risk as [age of nuclear power plants] × [number of past incidents] × [frequency of earthquakes typhoons/ human errors, etc.]. Having verified this disaster formula with the Fukushima Daiichi Plant accident data, we applied the index as calculated to the world's 441 nuclear facilities at the time of writing. Adjusting the calculations to match available information on such factors as reactor age, reported number of instances of failure, natural disasters, and risks of war, nuclear terrorism, and so on generated an indexical figure for each of the 441 facilities. Applying the value for overseas nuclear facilities by comparing it with the average figure for the Fukushima Daiichi Plant's six reactors (Ave. 141), one can indicate a statistical level of risk (Fig. 2.4).

Formula 2.1 Disaster Formula: age of plant, accident history, and precipitating factors (general formulation of organizational disasters) by the author's MEXT KAKENHI research group

$$\varepsilon = \alpha \times \beta \times \gamma / (\rho + \tau)$$

ε: disaster index
α: age/degree of degradation
β: frequency of instances of failure
γ: precipitating factors
ρ: confidence rating of organizational management
τ: credibility of nation state/management body

Regarding the measurement of the credibility level of social systems ($\rho + \tau$), the rating of organizational safety management and administrative supervision is an essential element in ensuring the safety of nuclear power. An index displaying industry organizations' and countries' or governments' credibility ratings along the lines of Moody's or Standard and Poor's 'credit rating' may be applicable. The credibility of the power industry can be established through their release of long-standing financial data to demonstrate their integrity, much as data about human health is obtained through a blood sample. In short, the system pathology of nuclear power can be comprehended as a multidimensional structural problem comprised of social, organizational, and economic elements.

Power companies' safety-management capacities and credibility ratings are linked: generally speaking, industry organizations' credibility level and safety management go hand in hand, in much the same way that a nation's credibility is evaluated from its financial rating as well as the market price for its bonds and currency. This is reflected in the level of investment in crisis-management capacities and safety measures against a possible disaster. Incidentally, since the 'Lehman shock' (the adverse economic effect of Lehman Brothers' collapse), it has become commonly accepted that, as with the Greek debt crisis in the EU, a nation's credibility rating can be calculated. Thus, we have incorporated these social systems' credibility into the Fukushima Formula to provisionally calculate the indexical value ε. In 2010, Fukushima Prefecture experienced four earthquakes rated '4' or higher on the Japanese scale of 1–7, and hence $\gamma = 4$ (for the four instances).

2.3.4 *Application of Disaster Formula: Global Nuclear Power Hazards*

Adding together the factors α (age), β (history of trouble), and γ (precipitating factors: frequency of earthquakes) for the Fukushima Daiichi Plant's six reactors, the average indexical value across the reactors is 141. However, even if sufficient safety management ρ and administrative direction from the state τ were being undertaken with regard to the aging reactors α, the number of instances of failure β and the frequency of earthquakes γ raise the probability of an accident. Even if, for argument's sake, the power companies' and State's administrative supervision is 100 % functional—that is, $\alpha/(\rho + \tau) = 1$—it is impossible to evade 'large-scale catastrophization' of an accident should the values representing the number of instances of failure β and frequency of earthquakes γ be high. This would apply equally to a newly constructed nuclear plant.

Table 2.4 indicates the calculations involved in arriving at an average value of 141 from the organizational-disaster formula for the Fukushima Daiichi Plant's reactors 1 through 6. The overall value for the safety management (ρ) and administration (τ) of the aging nuclear plant (α) was calculated as 1 ($\alpha/(\rho + \tau) = 1$),

Table 2.4 Risk indices of the Fukushima Daiichi Plant's six reactors

Indexical value ε of Fukushima Daiichi Plant's six reactors (Reactor 1 = ε^1, etc.)											
ε^1>216, ε^2>227, ε^3>124, ε^4>80, ε^5>84, ε^6>116, average = 141											
	ε	=	α	×	β	×	γ	/	(ρ+τ)		
Reactor 1	ε^1	=	40	×	54	×	4	/	—	≒	216
Reactor 2	ε^2	=	36	×	57	×	4	/	—	≒	228
Reactor 3	ε^3	=	35	×	31	×	4	/	—	≒	124
Reactor 4	ε^4	=	32	×	20	×	4	/	—	≒	80
Reactor 5	ε^5	=	32	×	21	×	4	/	—	≒	84
Reactor 6	ε^6	=	31	×	29	×	4	/	—	≒	116
Calculation based on α/(ρ+τ)=1										Avg.=	141

however, generally speaking, an aging plant is likely to have a higher danger value. Even if a power company and administrative authority claim to place the utmost importance on running a plant with, respectively, "perfect safety management and 100 % effective safety measures," the danger index will be high with an aging plant. This is because the accident risk comprising the number of instances of past trouble (β) and frequency of earthquakes (γ) is a decisive and unavoidable factor. The values outputted from the formula are the lowest-possible indices: thus, in cases where a power company and its safety-management system, or a state and its administrative-management system, are faulty, the danger level rises.

Applying the disaster formula to the world's 441 nuclear power reactors operational at the time of writing revealed many facilities exceeding the Fukushima Daiichi Plant's average index rating of 141. These are displayed in Fig. 2.4: World nuclear hazard map. Even if one sets the power company's credibility rating ρ (which is proportional to a plant's safety management) and the supervisory authority's level of credibility τ at the highest levels when calculating, the facilities exceed the Fukushima Daiichi Plant index rating. Prior to the Fukushima Plant accident, risk calculated from IAEA and World Nuclear Database data was largely in line with the dangers that media and Internet sources claimed to exist at nuclear facilities in other countries. It is of particular note that on the same day the World Hazard Map pictured above first appeared, in a presentation at an international systems studies conference at Hull University in the United Kingdom, *The Wall Street Journal* published in a special feature article its own US nuclear power hazard map by Nuclear Regulatory Commission (Fig. 2.5: US Geological Survey, *National Seismic Hazard Mapping Project*), which accorded with the World Hazard Map.

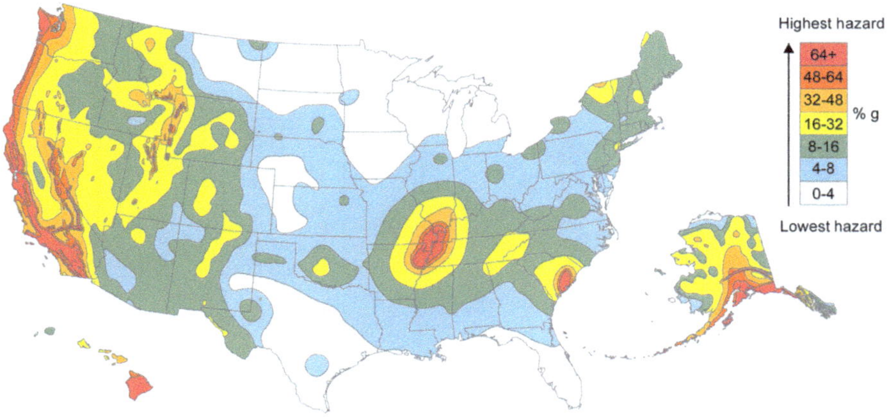

Fig. 2.5 US nuclear power hazard maps *Source:* US Geological Survey, National Seismic Hazard Mapping Project, Nuclear Regulatory Commission, 20 July 2011, pp.12–13 (*Released* on 21 July 2011 by Wall Street Journal))

2.4 Paradigm Shift to Sustainability

2.4.1 Renewable Energy and Electric Power Policy

Japan's rate of reliance on nuclear energy as a source of electricity has fluctuated between 25 and 30 %. The adoption of sources of so-called renewable energy such as solar, wind and geothermal power stalled due to regulations imposing a 1-percent upper limit on such renewables; these regulations have begun easing in recent years.

The historical background to Japan's dependency on nuclear power consists of the Japan—United States Atomic Energy Agreement, the Nuclear Nonproliferation Treaty, and beyond these the formation of the 'nuclear umbrella' of western and eastern ally countries. The transfer of nuclear-energy technology has been incorporated in the United States' global strategy. At the same time, the rapid rise in electricity demand in high-growth countries such as China, India, and Brazil is beginning to accelerate the promulgation of nuclear sources of electricity. Amid this, nuclear power is competitively priced compared with other forms of electricity generation such as thermal and hydroelectric, leading to the concomitant promulgation of 'safety myths' in which environmental risks and costs, including those to humans, are disregarded by unforeseen and beyond expectation disasters (Photograph 2.4).

However, just how low are the costs associated with nuclear power? Taking into account both the costs of reactor decommissioning and accident compensation, the recalculated generation cost provided by Prof. Kenichi Ōshima in 2010, prior to the Fukushima Daiichi Plant accident, is 10.68 yen/kWh, while that of thermal

Photograph 2.4 Kuji tsunami *Source*: by *permission* of Kuji city office authority

Table 2.5 Provisional list of nuclear power electricity-generation cost calculations

	Institute of Energy Economics, Japan August 2011	Research Institute of Innovative Technology for the Earth May 2011	Energy White Paper 2010 June 2010	Kenichi Ōshima
Nuclear	7.2	8 – 13	5 – 6	10.68
Thermal	10.2	(coal) 5 – 7 (LNG combined) 10 – 14	(LNG) 7 – 8	9.9
Solar PV	–	55 – 63	49	–
Geothermal	(Mainly geothermal) 8.9	–	8 – 22	–

(Unit: Yen/kWh)
Source: Based on K. Ōshima and website (http://www.nexyzbb.ne.jp/~omnika/hatsudentanka.html)

generation is 9.9 yen/kWh, making nuclear power provisionally more expensive (Table 2.5) [19].

Provisional list of nuclear power electricity-generation cost calculations shows the generation costs released after the accident at the Fukushima Daiichi Plant employ data produced by such research bodies as power companies' funds that do not incorporate environmental risks or costs and thus fail to reflect the public's viewpoint. Reactor decommissioning costs at the Chernobyl Plant were similarly not covered by the national budgets of Ukraine or Belarus, meaning there was no guaranteed compensation for accident cleanup or local residents' needs. When one compares the cost findings of Prof. Ōshima and the Research Institute of Innovative

Technology for the Earth (RITE) and those of other bodies, it is clear that estimates for nuclear power that fail to take into account ecological and social fundamentals like Spaceship Earth—human life and habitation—as well as farmland and fisheries can by no means be called an 'economical and commercial' option.

Traversing the international landscape of nuclear power, one finds that among the world's some 441 reactors, the United States possesses the most with 104, followed by France with 59, and then Japan with 54 reactors. However, in response to the Fukushima Daiichi Plant accident, Germany, Italy and Switzerland have led a shift toward denuclearization. On the other hand, China plans to build additional large-scale reactors up to a total of 66 by 2030, and India has announced new, albeit small-scale, nuclear power projects, 25 years after the Chernobyl accident. The expansion of nuclear power in electricity-hungry nations like China, India, and Brazil is playing a part in the energy strategy of a virtual nuclear energy 'conglomerate' consisting of the United States, through their nuclear energy business and nuclear security (military affairs), and France and Russia in terms of nuclear-waste business. At an international conference on Systems Sciences in 2011, some policy scientists memorably characterized how the United States controls the military, political and economic fundamentals of global society through a "nuclear weapon and energy business strategy" that puts the United States at its core.

2.4.2 Potential for Nuclear Power Phaseout

In 2011, the WWF (Worldwide Fund for Nature) released a report aspiring to the target of realizing 100-percent renewable energy by 2050 [20]. Former Japanese Prime Minister Hatoyama had previously announced a target of a 25-percent reduction in CO_2 levels by 2020 at the United Nations General Assembly, but this has continued to be revisited in the wake of the Fukushima Daiichi Plant disaster, and energy-saving measures were put in place after the devastating earthquake. Energy-saving initiatives subsequent to the nuclear accident, such as installation of LED (light-emitting diode) lamps and ICT (information and communication technologies), as well as energy-efficient 'smart grids,' have meant a burgeoning business in power saving.

Since power consumption fluctuates according to season and time of day, exploration has begun of social-systemic initiatives such as shifting to night operations at factories and other sites so as to utilize surplus electricity at those times. *Inter alia*, the work style that is a historical and cultural artifact of postwar Japan's period of rapid economic growth is being diversified. It already reflects the modern diverse lifestyle exemplified by convenience stores and places to eat out open 24 hours a day. Energy scenarios (Fig. 2.6) demonstrate that under the present circumstances in Japan, electricity-transmission loss rates are high, with only 50 % of the total amount of energy produced being actually delivered. No matter how much energy nuclear plants produce, a large percentage of the electricity is lost

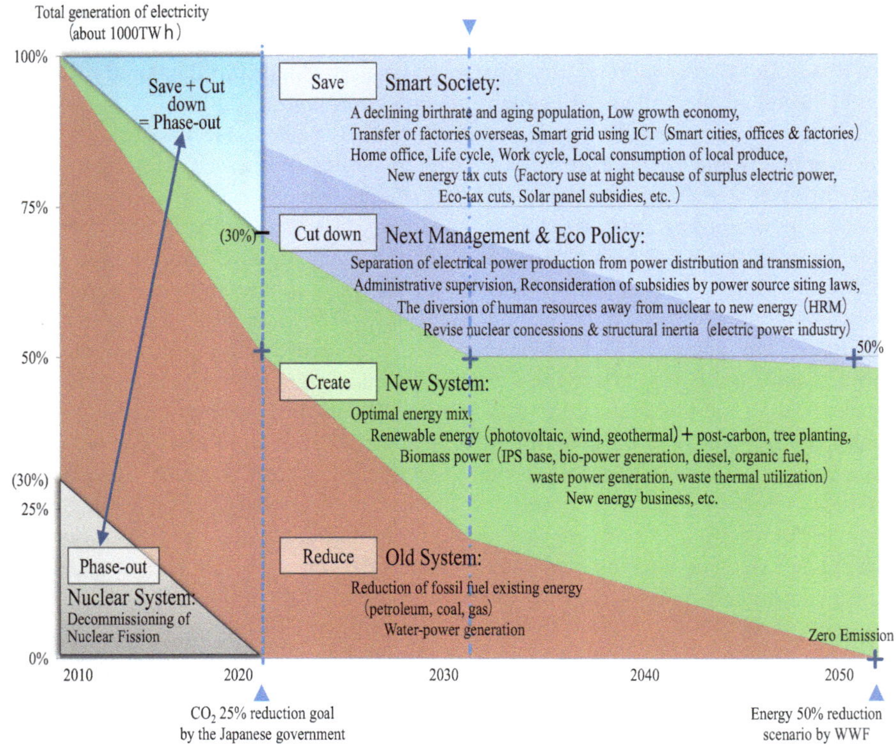

Fig. 2.6 Energy scenarios according to author's seminar by MEXT KAKENHI. *Note:* 'Cut down' inspired by K. Hioki of Kyoto University and T. Tokugawa, President of WWF Japan

through such transmission attenuation. While it is ideal to have electricity production that responds to demand in a timely fashion, the loss inherent in the physical distances between the places where electricity is produced and used cannot be ignored.

Postscript: During a presentation, Prof. K. Hioki identified the need for those vested-interest groups surrounding electricity production to eliminate 'wastage' before next-generation energy could be considered. From this point, we came to the conclusion that in addition to pursuing 'power saving' on the demand side, it was also necessary to redress the 50 % transmission loss on the supply side occurring between production and actual use of electricity. The supply side (power companies and electricity policies) has an obligation to work toward the decoupling of electricity production and delivery; reduced attenuation of electricity; and local production/local consumption. As a foil to the call for power saving among all individual citizens, we have labeled the efforts of the power industry and administrative supervision themselves 'electricity conservation.'

Obviating this 50-percent loss in transmission and distribution through the "local production and local consumption" of electricity would comfortably cover the 25–30-percent reliance on nuclear power, making it possible to consider eliminating atomic energy entirely. At the same time, however, 'energy trading' whereby power companies buy private individuals' surplus production of 'clean and green' energy may eventually have a negative impact on consumers' and users' electricity bills. The reason for this is that we, private electricity consumers, may be landed with the bills from new electricity-generation entities even as they make profits, having passed the barriers to entry amid the relaxation of regulations in the electricity industry and introduction of compulsory energy trading. Such a result would not solve the fundamental problems, and a vicious circle of investment in power-saving facilities due to the nuclear accident would continue its cycle, with the very entities that contributed to the problem being enriched in the process of addressing it.

Policy decisions toward the liberalization of electricity must redress the structural inertia of the vested interests in the electricity industry, and policy shifts leading to the decoupling of electricity generation and distribution, along with local production/local consumption, should curb electricity loss. While such liberalization should contribute to reducing the burden on private individuals, at present it is actually regressing. If one observes 'social publicness' from a macro perspective, electricity conservation as it relates to power saving, production, and reduction has an even higher degree of efficacy. The catastrophization of the nuclear accident served to clarify the existence of the various stakeholders behind the scenes in the electricity realm, and sustainable energy policy-making has become an urgent task.

The vested-interest structure of the electricity industry and their call for individuals to save power, were both partly brought to light by the nuclear accident, revealing the system pathology of Japan's social organization. Strangely, the Fukushima Daiichi Plant disaster traced the same locus as the 1999 criticality accident, as though accumulating on top of wartime and historical disasters. In future, with the shift from the structural inertia of the electricity industry and nuclear power administration into new energy business, the creation of new forms of employment will be an urgent issue for the social and economic systems, and efficiency in human resources and the production and distribution of electricity will likely be indispensable in the energy field. At the same time, during a period of low economic growth, itself within an era of low birthrate and an aging population in Japan, there will be an inevitable hollowing-out of industry, with the relocation of factories overseas and an accompanying 'brain drain,' rendering uncertain the future electricity demand [21]. The denuclearization trend is probably incompatible with the direction the financial and industrial worlds are taking. On the other hand, with a shrinking population due to the low birthrate along with the hollowing-out of industry and the economy, overestimating electricity demand and continuing to operate aging nuclear power plants can be characterized as potentially running against the trend toward a sustainable society in the near future. The scenario is synchronized with the 2020 goal of 25 % reduction in CO_2 by the Japanese government and the WWF's goal of 'zero emissions' by 2050.

The nuclear disaster under consideration dispossessed those inhabitants within a 20 to 30 km radius of the most fundamental components of their everyday lives—such things as their health and property (homes, farmland, and fisheries). It is possible too that restarting aging nuclear reactors at this time is an unlawful act. Just as with the Fukushima Daiichi Plant itself, one can say that the technology, organizational, and policy systems designed during the period of rapid economic growth have manifested structural aging. This has stemmed from skepticism concerning Japan's internationally vaunted technological systems, exposing an insoluble system-wide dilemma of organizational management with inadequate safety control, and administrative supervision and electricity policies dominated by structural inertia. In the present chapter, we have attempted to provide an overview of the individual, organizational, and policy elements and to present a next-generation sustainable-energy scenario as a paradigm for electricity production in regard to such system-wide deterioration and aging.

Figure 2.6 outlines a shift in the sources of electrical energy whereby the optimization of power saving and electricity conservation may render nuclear power unnecessary. Further, it illustrates the great potential for the new energy 'best mix' to produce enough energy to compensate for the corresponding reduction in fossil-fuel energy production.

2.5 Decision-Making for Sustainability

Having experienced the unprecedented disaster of the Fukushima Daiichi Plant, the trend is for individuals to place relatively more importance on safety issues. At the same time, distrust of the siting of nuclear plants (an issue of plant administration) and their safety management (an issue of power-industry management) is growing. Electricity production is at a societal and economic crossroads as to whether or not nuclear power will continue to be chosen as an energy source in the future. The sustainability of a sense of safety is embedded in what H. A. Simon calls 'factual and valuable premises' [22] that impinge on our subconscious. The terror and menace embodied in the nuclear accident realized themselves at the level of the everyday, and the experience with its accompanying sensations provoked a desire for safety among the general populace. The propagation of awareness of such a decisive prerequisite brings into question the decision-making around the restarting of nuclear reactors, and restores to its proper place the executive power that we had handed over to the State or 'the fiction of superior authority' [23].

Sustainable decision-making around sources of electricity (Fig. 2.7) shows the shift from a two-dimensional balance of supply and demand on an *x*–*y* axis in which generation costs only incorporate the minimum of economic-related decisions, to a three-dimensional decision structure that also incorporates environmental restoration costs (the *z* axis). For example, the large number of people who have massed in front of the Prime Minister's official residence in Tokyo on a weekly basis demanding denuclearization has been an enduring rather than transitory display of

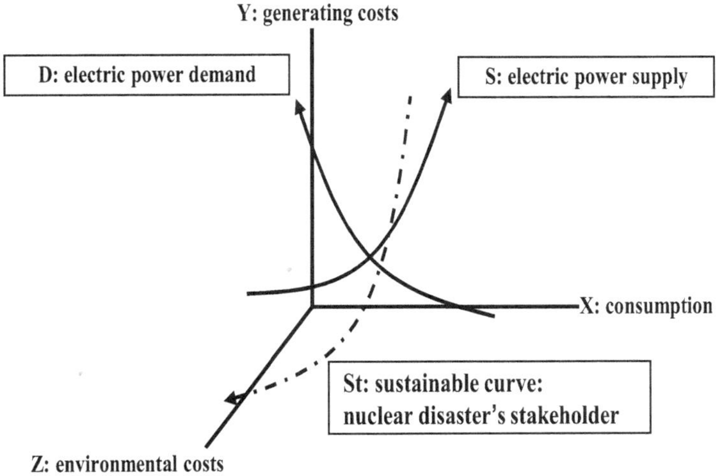

(for restoration of ecosystems damaged by radioactivity, including human society in the environment)

Stakeholders: Seeking restoration of radiation area, recovery of local people's health, agriculture, and contaminated ocean.

Fig. 2.7 Sustainable decision-making around sources of electricity *Note:* drawing a diagram based on the concept of 'sustainable curve' at ISSS workshop by A. Laszlo 2011

individual action expanding across the country that has shaken the policy decisions around restarting the reactors. If the collective decision-making of the stakeholders who guarantee safety and seek sustainability becomes the majority, then it will bear some of the functions of the social system. The decision process regarding nuclear power is multidimensional, comprising economic, political, and organizational elements whose effects transect society.

In countries such as Germany, Italy, and Switzerland, citizen's selection criteria for energy sources place more emphasis on societal and safety factors than on those relating to the economy or efficiency. When such thinking is in the majority, it fosters a social rationale that is reflected in decision-making and actions. G. E. Swanson suggested that, in particular, the rationale formed under severe circumstances like a nuclear disaster is superior to previous economic or even 'scientific rationality' [24]. The nuclear question shakes people, whether individuals or members of organizations, at the core of their unconscious, becoming a turning point in the balance between economic and social rationales.

An illustration of the abovementioned points is how the housewives in Date City, Fukushima Prefecture, in the middle of the nuclear disaster, acted to protect their children by investigating radioactivity contamination levels and demanding that the administration distribute dosimeters to every resident. Such an act of 'species preservation' *Kizuna* based on love was surely one part of the emotion-based group consciousness in response to fear of the intangible, invisible, and odorless radiation that mobilized the regional administration. The crisis situation mobilized the cultural system of mutual assistance that placed warm human relations at its center. Thus it can be said that the damage from the earthquake and tsunami manifested the societal, economic, and organizational systems latent in human mutuality. This is reminiscent of H. R. Maturana's statement that "emotion is a programmed survival mechanism for the preservation of the species" [25].

The Fukushima Daiichi Plant accident began with reactor-core meltdown due to insufficient safety management, leading to large-scale release of radiation: A natural disaster evolved into an accident and then a human-made disaster. The catastrophization process of the accident exposed the structural error inherent in the integrated system of nuclear technology, safety management, nuclear administration, and electricity policy, but on the other hand signs of recovery are to be found around the societal and cultural systems. In this chapter we have attempted to interrogate the paradoxical system pathology that has inevitably arisen from the historical fundamentals of Japan's societal, economic, and management (organizational) systems. It appears that a new era of vigilance is upon us in which individuals who worry about radiation levels in their food are left to fend for themselves.

References

1. Atsuji, S. et al. "The Fukushima Catastrophe seen as a Malfunction of Organizational Systems: Nuclear Fade-out and Hazard Maps for the World", *General Systems Bulletin,* Vol. XXXXI, 2012, pp. 26–30.
2. Thom, R., Zeeman, E.C., Ujiki, S. and Sawa, T., *Keitai to kouzō—katasutrofu no riron* [*Catastrophe Theory*], Misuzu Shobō, 1977.
3. Yablokov, A.V., Nesterenko, V.B., and Nesterenko, A.V., *Chernobyl: Consequences of the Catastrophe for People and the Environment*, Blackwell, 2009.
4. Troncale, L.R., "Comparing Systems Pathology Treatments Across Systems Processes Theory", *Proceedings of the 55th Annual Meeting of the ISSS*, July 2011, p. 115.
5. Masai, Y., (ed.), *Ima ga wakaru jidai ga wakaru nihon chizu* [*Understanding the Present, Understanding the Period: Map of Japan*], 2010 edition, Seibido Shuppan, 2009, p. 93.
6. Nuclear and Industrial Safety Agency, "Genshiryokuhatsudensho no hoan katsudō hyōka (shikō) no kekka ni tsuite" [Regarding (trial) evaluation of activities to maintain nuclear plants], June 2010.
7. Japan Nuclear Energy Safety Organization, "Jishin reberu 2PSA no kaiseki (BWR)" [Level 2PSA Earthquake Analysis], October 2010, pp. 4–7.
8. Saitō, H., "Fukushima genpatsu wa jishin ni taerareru no ka", in *Genpatsu wa jishin ni taerareru ka* [*Can Nuclear Plants Withstand an Earthquake?*], Citizen's Nuclear Information Center, 2008, pp. 29–33.
9. *Nihon Kōgyō Shimbun* publication, *Kakusareta genpatsu deeta* [*Concealed Nuclear-plant Data*], Fuji Sankei Group Nihon Kōgyō Shimbunsha, 2003, p. 14.
10. Citizen's Nuclear Information Center, *Kenshō Tōdengenpatsu toraburugakushi* [*Inspection: Concealment of Trouble at Tokyo Electric's Nuclear Plants*], Iwanami Booklet no. 582, 2002.
11. Atsuji, S. and Senba, S., "Denkijigyōjiyūka to shakaikōkyōsei", in *Kōgyō keiei kenkyū* [*Industrial Management Research*], Kōgyō Keiei Kenkyū Gakkai, 2002, vol. 16, pp. 22–26.
12. Tōkyō Denryoku Fukushima Genshiryoku Hatsudensho Jiko Chōsa Iinkai [Tokyo Electric Fukushima Nuclear Plant Accident Commission], *Kokkai jiko chōhōkokusho* [*Accident Report to the Diet*] *See also*, Fukushima Genpatsu Jiko Dokuritsu Kenshō Iinkai [Fukushima Nuclear Plant Accident Independent Inspection Committee, *Chōsa/kenshō hōkokusho* [*Investigation/Inspection Report*].

13. Chroust, G., "Regional Disasters and Systemic Reactions", *Proceedings of the 55th Annual Meeting of the ISSS 2011*, July 2011, p. 97.
14. Reason, J., *Managing the Risks of Organizational Accidents*, Ashgate Publishing, 1997, pp. 11–17.
15. Reason, J., *The Human Contribution: Unsafe Acts, Accidents, and Heroic Recoveries*, Ashgate Publishing, 2008.
16. Barnard, C.I., *The Functions of the Executive*, Harvard University Press, 1938, pp. 25–41.
17. Kemeny, J., *The Need for Change: The Legacy of TMI. Report of the Presidential Commission at Three Mile Island*, New York: Pergamon, 1979.
18. Reason, J., "The Chernobyl errors", *Bulletin of the British Psychological Society*, Vol. 40, 1987, pp. 2201–6. *See also*, Yablokov, A.V., Nesterenko, V.B., and Nesterenko, A.V., *op. cit.*
19. Ōshima, K., *Saiseikanō enerugii no seijikeizaigaku* [*Political Economy of Renewable Energy*], Tōyō Keizai Shinpōsha, 2010, p. 080.
20. WWF, *The Energy Report—100% Renewable Energy by 2050*, released November 10, 2011.
21. Hioki, K and Nakamaki, H., *Keiei Jinruigaku Kotohajime [Beginning; Management Anthropogy]*, Touho, 1997.
22. Simon, H.A., *Administrative Behaviour*, 3rd edition, The Free Press, 1976.
23. Barnard, C.I., *op. cit.*, pp. 170–171.
24. Swanson, G.E., "Management Observation and Communication Theory", *Proceedings of the 44th Annual Meeting of the ISSS*, July 2000, p. 121.
25. Maturana, H.R., "Ethics as Human Action—Ethics Belongs to the Emotional Domain", *Proceedings of the 44th Annual Meeting of the ISSS*, July 2000, p. 91.

Chapter 3
Our Stolen Sustainability: Contamination by Environmental Hormones

When we speak of sustainable development in the context of modern society, what does it actually mean? Here we would like to address the issue of environmental contamination and explore the conditions required to move toward human coexistence and symbiosis with the natural environment [1]. Studying the effect of 'environmental hormones' [2] and endocrine disruptors on the ecosystem, Theo Colborn confirmed the risk of 'brain contamination' through the food chain and 'biomagnification'. The contamination of the ecosphere by environmental hormones—an issue which no modern human can avoid—is a 'negative legacy' contrasting with the economic wealth brought by modern rationality. In this chapter, I consider the role of the environmental ethics which societies and individuals will be required to practice in the near future and examine the role of 'environmental management' as a preventive policy informing the morality of modern organizations.

3.1 Brain Contamination by Environmental Hormones

3.1.1 Biomagnification of Environmental Hormones

The endocrine disruptors which are known collectively as environmental hormones are recognized to include around 600 substances, all of which are artificially compounded chemical substances. The best known are dioxins and polychlorinated biphenyls (PCBs). Even in extremely small amounts, these substances disturb the endocrine system that controls the body's internal regulatory functions.

Reworking of: Atsuji, S., "ISO Global Policy toward Environmental Management", *Economic & Political Studies Series No. 135*, The Institute of Economic and Political Studies, 2003, pp. 1–18.

S. Atsuji, *Unsafety*, Translational Systems Sciences 7,
DOI 10.1007/978-4-431-55924-5_3

Environmental hormones are also contained in dichlorodiphenyltrichloroethane (DDT) and other agricultural chemicals, herbicides, and plasticizers, and in the artificial preservatives, colorants, and flavorings used in processed foodstuffs. In recent years, atrazine and bisphenol, which are contained in plasticizers, have been found to affect sex determination in all animal species and to alter the male–female birth ratio. Unbalanced sex determination threatens the preservation and the very survival of the species. Instead of being broken down, environmental hormones—chemical substances not originally present in the natural world—accumulate in the body. As a result, these artificial chemical compounds unleash a chain of contamination through the bio-environments which are formed through the interdependence of diverse lifeforms.

As environmental hormones cannot be detoxified or excreted by the organism itself, serial accumulation takes place over the long term in proportion to the amount ingested. Focusing specifically on the harmful substances of the organochlorine family, these were previously not considered a threat because, after diffusing through the soil in the form of agricultural chemicals and herbicides, they were thought to be rendered safe by a process of deposition in rivers and lakes via rainwater and groundwater drainage and soil permeation. The truth, however, is that these harmful substances, far from being removed, continue to circulate in the ecosystem through the food chain until they are served back up, in concentrated form, to the humans who stand at the top of the chain. In her work *Our Stolen Future*, T. Colborn explains this mechanism, which she calls biomagnification [3].

As illustrated in Fig. 3.1, T. Colborn indicated 'accumulative contamination with environmental hormones' has already spread to many animal species through the food chain. The harmful chemical compounds mentioned above are raised from river and lake sediment to the surface layer by plankton photosynthesis and predation. Contaminants which were previously assumed to sink to the river or lake bed are in fact ingested by the interconnected food chain of living organisms stretching from phytoplankton through zooplankton, immature fish, and adult fish to birds, becoming successively concentrated in their bodies instead of being excreted. This startling fact was demonstrated by T. Colborn using data from a large number of researchers. Environmental hormones inflict serious damage on the reproductive-cell DNA of all animal species in the global ecosystem and threaten their very survival. PCB, DDT, and other agricultural chemicals were thought to be deposited in the sedimentary layers of the hydrosphere and in the soil; in fact, through the food chain and biomagnification they come hurtling back to humankind like a boomerang.

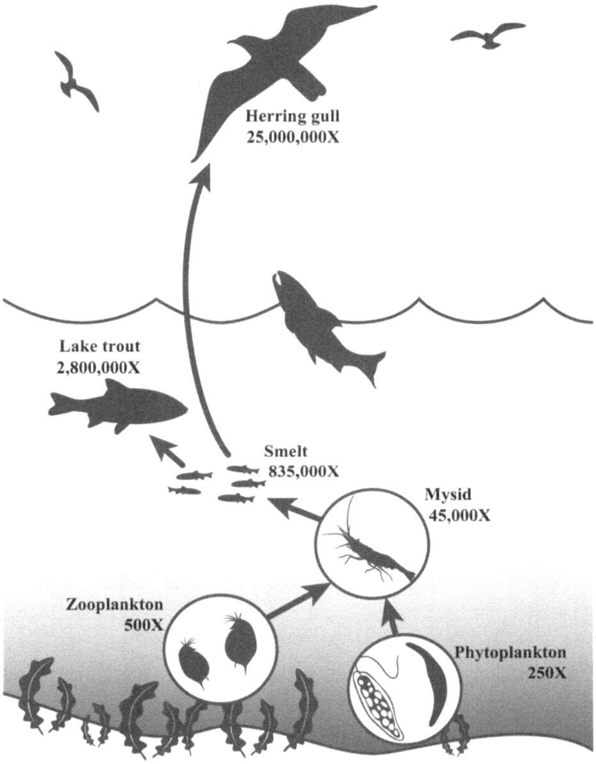

Fig. 3.1 Biomagnification of PCBs (bio-accumulation in the food chain). *Note:* Illustration based on Colborn et al. [2], p. 27

3.1.2 Brain Contamination through Environmental Hormones

According to T. Colborn, 'brain contamination' [4] is the most significant effect of environmental hormones. Dioxins and PCBs accumulate in the mother's body and impair the formation of the cranial nerves during the fetal growth stage (Fig. 3.2). In a study of the Great Lakes, J. L. Jacobson and S. W. Jacobson demonstrated that PCB impedes the formation of the fetal brain in the womb [5].

In the Jacobsons' studies, children exposed to PCB at the fetal stage were observed to suffer damage to brain function, with intellectual impairments including reduced memory capacity and decreased environmental adaptability. Brain contamination not only has a general disruptive effect on human physiological function but also causes mental impairment and irreparable pathologies. Accordingly, T. Colborn posited an effect of environmental hormones and other contaminants on the children of the next generation. P. Hauser asserted that the sharp

COUNTRY	1987/88	1992/93
Thailand	4.9	
New Zealand	5.8	
India	6	
Hungary	10.2	8.2
Croatia	11.9	11
USA	16.6	
Finland	16.8	16.8
Norway	17.8	10.6
Denmark	17.8	15.2
Austria	17.9	10.8
Vietnam	18.3	
Canada	18.5	14.5
Poland	20.8	
Sweden	22	
Japan	23.9	
Germany	31.8	16.5
UK	33.1	16.6
Netherlands	37.1	22.4
Belgium	37.6	24.8
	(TEQpg)	(TEQpg)

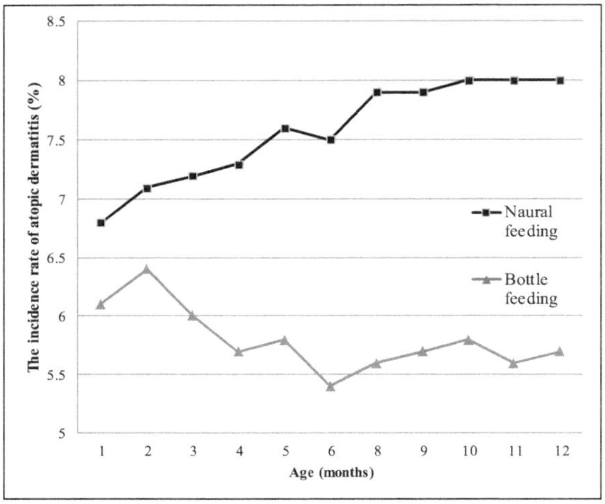

Fig. 3.2 WHO research on breastmilk contamination. WHO research on incidence of atopic dermatitis. *Source:* H. Hongo, *Iwanami Booklet No. 482*, *Bonyū to Dioxin*, Iwanami, 1999

increase in emotional disorders among children in the form of suicide, bipolar disorder, uncontrollable anger, and attention deficit hyperactivity disorder (ADHD) [6] was a symptom of this whose origins did not lie solely in the unhealthy social environment of today; damage to brain function by environmental hormones was also a factor. The correlation between brain damage through ingestion of chemical substances and abnormal behavior has already been proven through experiments in rats [7].

It is indeed impossible to rule out a link between environmental hormones and the current sharp rise in the suicide rate, the increased incidence of depression, and the emergence of modern illnesses such as ADHD. As the mechanism of the brain damage which is thus able to paralyze the nervous system is gradually elucidated, there is concern that functional impairment during the developmental process due to brain contamination may derange people's judgment, and that the resulting abnormal behavior, played out on a larger scale, may inflict major damage at the level of the organization or the state. Colborn maintained that the biggest threat from environmental hormones lay in brain contamination, which could bring not only contamination of the bio-environment but also destruction of the social environment, and was concerned over the impact of organizational behavior on the environment. Going forward, she predicted, if people affected by brain contamination became the majority, this would have a considerable impact on state policy and organizational decision-making and eventually have a direct and decisive influence on the survival of humankind. From this it can be appreciated that resolving the issue of environmental hormones is an urgent task not only in terms of the natural bio-environment but also the social environment.

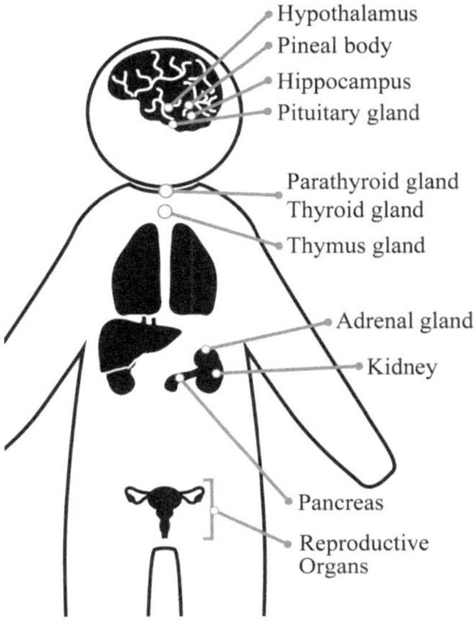

Fig. 3.3 Glands, organs, and tissues sending or receiving hormonal messages in the human body. *Note:* Illustrating based on Colborn et al. [2], p. 33

It is 50 years since Rachel Carson sounded the alarm over the dangers of insecticides, agricultural chemicals, and also pharmaceuticals such as antibiotics in her book *Silent Spring* [8]. Carson's predictions were accurate. Various chemical substances have been demonstrated to contaminate animal species and the ecosystem and to accumulate in the human body. Created by artificial manipulation, insecticides and other synthetic chemical substances are not detoxified after release into the natural world but rather spread their contamination, causing a grave situation. Colborn's work served almost as a systematic verification of Carson's predictions and revealed the possibility of brain contamination that could cause modern humans to take deranged decisions and actions.

Figure 3.3 shows the mechanism whereby environmental hormones act on thyroid hormones and damage the regulatory functions of the human body. There is particular concern over possible fetal brain contamination. T. Colborn also confronts us with the terrible danger posed by the chemical by-products of today's consumer society. Her fear was that the production volume of synthetic chemical substances was positively correlated with the prevalence of self-destructive mental states.

3.1.3 The Globalization of Environmental Contamination

Environmental contamination began as a problem of developed countries with highly advanced societies, but as contamination has spread, its range has come to extend over the whole of the global ecosystem. Even though its sources may be

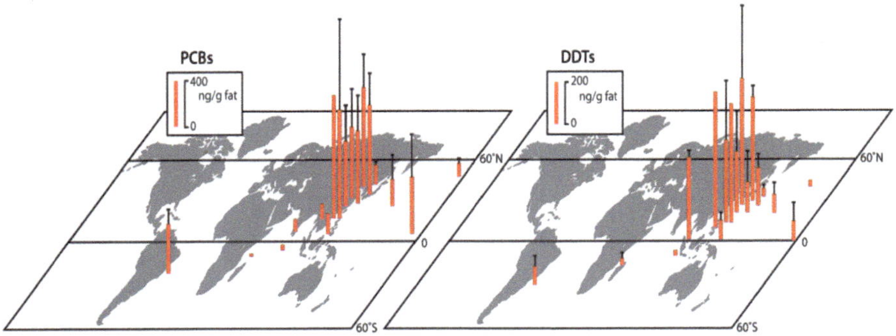

Fig. 3.4 PCB and DDT contamination in liver of skipjack tuna. *Source:* WHO Report, State of the Science of Endocrine Disrupting Chemicals—2012, p. 212

located in a specific area, contamination becomes dispersed through convection in the atmosphere and hydrosphere. Reinforced regulation has the effect of turning the industrial effluent, exhaust gases, and other emissions and industrial wastes of developed countries into pollution for overseas export to developing countries, while emission credits and other forms of atmospheric trading are creating a seamless distribution of environmental contamination throughout the ecosphere. Similar is the use of chlorofluorocarbon gases, which rise to the stratosphere, where they create an ozone hole through which cosmic radiation reaches the earth's surface and impacts on animal species. This phenomenon is evident in many places throughout the world, and is spread worldwide through atmospheric and hydrospheric convection. Environmental contamination itself is thus being globalized [9].

Figure 3.4 shows levels of PCB and DDT in skipjack-tuna liver. Measurements are from the Pacific, Indian, and Southwest Atlantic oceans and do not cover the North Atlantic. There can be no such thing as locally limited environmental contamination. Convection through the water and air that make up the global ecosphere and interactions due to the movements of animal species such as migratory birds and fish that move with sea currents have the effect of disseminating contamination, so that environmental destruction spreads throughout the ecosphere. This may initially involve contamination at low concentration, but chemicals compounding with other substances and reactions with various chemical elements lead to the secondary development of toxicity and biomagnification in a 'domino effect of environmental destruction'. Environmental contamination can no longer be a distant concern, but is casting an unmistakable shadow on the lives of modern humans. Environmental hormones have a multiple presence in our lives, not only in synthetic preservatives, colorants, and flavorings, as well as herbicides and agricultural chemicals, but also for instance in dioxins emitted through waste incineration, and these substances have left residues in the bodies of nearly every human being.

Environmental hormones	Substances
(1) Persistent organohalogens	Dioxins, furans, PCBs, PBBs
(2) Pesticides	DDT, DDE, Atrazine
(3) Penta- to nonyl-phenols	Penta- to nonyl-phenols
(4) Bisphenol A	Bisphenol A
(5) Phthalates	DEHP, BBP, DBP, DPP, DHP
(6) Styrene dimers and trimers	Styrene dimers and trimers
(7) Benzopyrene	Benzopyrene
(8) Heavy metals	Cadmium, lead, mercury
(9) Other chemical bonds	2, 4-dichlorophenol, benzophenone

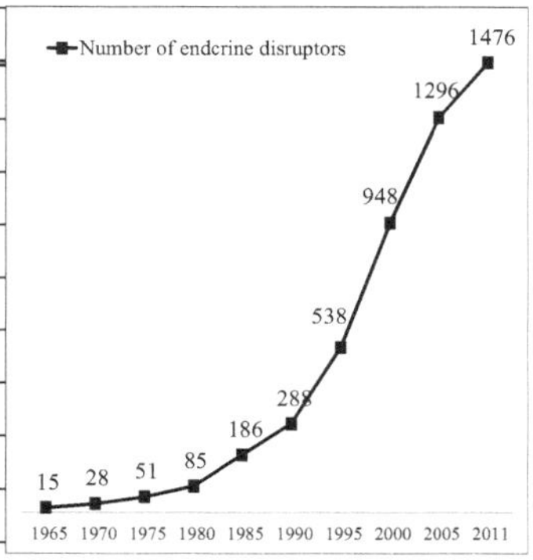

Fig. 3.5 Classification of environmental hormones and number of endocrine disruptors. *Note:* Extracted from TEDX (The Endocrine Disruption Exchange) and WWF releases

The threat from environmental hormones highlighted by T. Colborn was that, mediated through the interaction between living organisms that takes place in the food chain, contaminants would undergo biomagnification and collect in high concentrations in the bodies of modern humans. This would mean damage to DNA, which is a veritable 'blueprint for preservation of the species'. The contamination of animal wombs with environmental hormones amounts to the contamination of the global ecosphere, which is, as it were, the amniotic fluid of all life on earth—in other words, contamination of Gaia. This is because contamination of any one group of individuals, such as waterfowl, poses a danger to all living organisms, and in particular all intelligent lifeforms with a nervous system dependent on water and carbon compounds. Waterfowl contamination leads ultimately to 'Gaia contamination', or contamination of the whole of the global ecosphere, which threatens human survival. According to TEDX (founded by T. Colborn), the number of endocrine-disruptor chemicals suspected of acting as environmental hormones (*see* classification in Fig. 3.5) was in excess of 1,400 as of 2011.

3.2 Management for Environmental Protection

3.2.1 Global-Scale Spread of Contamination

It has already been reported that alligators in the US state of Florida have been found to display 'symptoms of sexual involution', with many males observed to have atrophied sperm glands. Physical analysis of the alligators showed a correlation between chemical concentration and sexual involution [10]. The concern here is the possible effect on humans. Since the alligator and human endocrine systems are similar, there will inevitably be a similar effect on humans. This appears to be backed up by survey reports in recent years indicating that the sperm count of adult

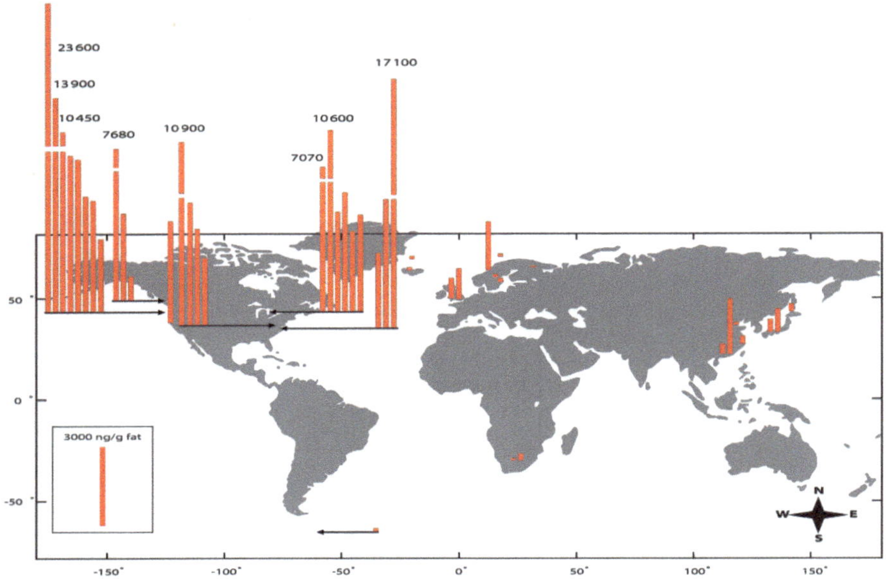

Fig. 3.6 PBDE contaminations in bird tissues and eggs. *Source:* WHO Report, State of the Science of Endocrine Disrupting Chemicals—2012, p. 214

males has halved. The endocrine disruptors nonylphenol and bisphenol destroy the endocrine receptors of fish species in contaminated water. Especially since animals move around, the area of secondary contamination gradually expands. With nonylphenol, movement on the part of the contamination source has the effect not of reducing the level of toxicity, but of extending the contamination range through the interaction between living creatures in the food chain. Characteristic of environmental hormones is global spread through the movement of species to reach seamless distribution. For instance, many of the waterfowl that inhabit Lake Ontario undertake seasonal migration to countries across the globe; behavior through which the contamination source is itself spread. As is already known, since almost none of the toxicity is removed, the biomagnification of environmental hormones intensifies progressively.

Figure 3.6 shows contamination with PBDE (poly-brominated diphenyl ethers) among birds in various regions, which is correlated with PBDE production levels at chemical plants. As PBDE has a similar structure to PCBs and dioxins, it has a high degree of lipid solubility, has accumulative potential, and is subject to bio-concentration.

The survey finding that environmental hormones have already been detected in 80 % of aquatic animals indicates that environmental contamination is not a transitory phenomenon restricted to single geographic areas. Today, moreover, there is concern over potential effects on unborn children due to the increased number of births to older women and contamination of breast-milk with PCBs. Do we have no choice but to carry on living with environmental hormones accumulating in our bodies?

The accompanying destruction of the global environment is also a serious concern. Most of the oxygen in the atmosphere (75 % according to surveys) is said to be dependent on the tropical rain forest of the Amazon basin. The felling of this forest coupled with slash-and-burn farming are destroying its ecosystems, from the trees which have taken thousands of years to grow to the fungi and bacteria in the soil, and are causing the loss of the forest. Similar deforestation is apparent everywhere around the world, and it is well known that acid rain caused by CO_2 emissions is stripping away the forest of Germany and other areas in Europe and the northern hemisphere [11]. Specifically, there is an area of Eastern Europe known as the Black Triangle [12], which remained hidden from the western media, where the environmental destruction was at its gravest, but it has recently become clear that serious atmospheric contamination and forest destruction are in progress here. Deforestation not only causes oxygen deficiency. When tree dieback spreads over wide areas, the water-retaining properties of mountainsides are weakened and they become unable to store rainwater, which washes into rivers and lakes and is carried directly to the sea. The bleeding of water resources from the land leads to progressive global desertification and the loss of the equilibrium between atmosphere, soil, and hydrosphere [13]. Environmental contamination and the destruction of the ecosphere thus aggravate each other in a synergistic fashion.

Figure 3.7 shows predicted distribution of PM2.5 atmospheric pollution according to the Japan Weather Association. PM2.5 refers to the size (2.5 μm or below) of micro-particular pollutant substances; the finer the particles, the more

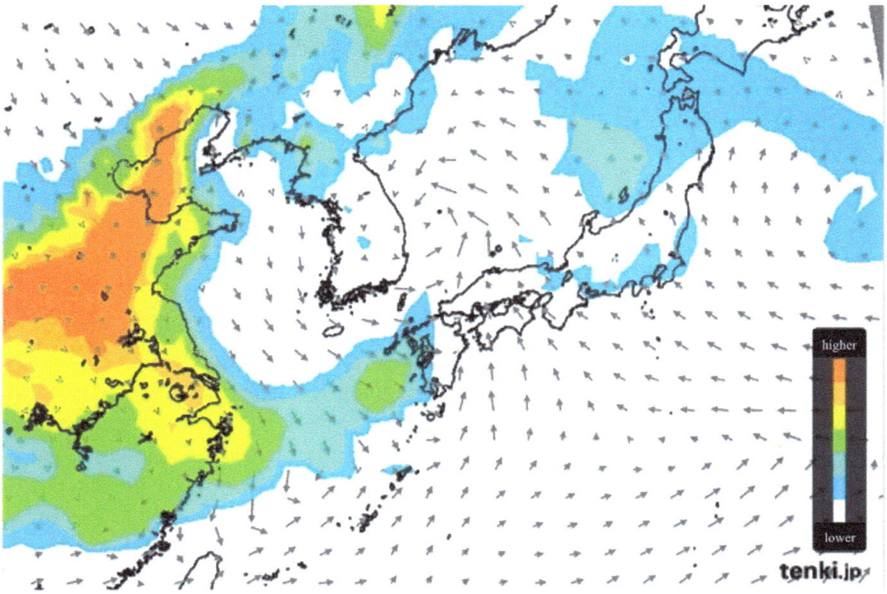

Fig. 3.7 Predictive map for distribution of PM2.5 (May 19, 2013), by *permission* of JWA. *Source:* Japan Weather Association URL: http://www.tenki.jp/

easily bio-accumulation takes place, leading to concern among experts over health damage.

Add to this situation the population explosion in China and India, and we face not merely a food crisis, but an era in which even the supply of air and water cannot be guaranteed. The activities of modern humans have ended up unwittingly creating for the children of the next generation a savagely hostile bio-environment and social environment.

3.2.2 Policy for Environmental Protection: Learning Lessons from Failure

When T. Colborn sounded the alarm over environmental hormones, she was referring to contamination not only of the bio-environment but extending to the social environment, including human psychology, and linked it further to the spiritual destruction of the individual. The physical, biological, and social impact of environmental contamination is connected with the survival and prosperity of humankind. Contamination of the bio-environment and destruction of the social environment show the heavy involvement of actions at the organizational and social rather than the individual level. Implicit in the consumer society of mass production and mass distribution operated by modern corporations are constituent factors that carry the seeds of effective environmental destruction. The direction for environmental protection to take going forward is shifting, within both corporate decision-making and government policy formulation, from a decision made at the individual level to one at the organizational and in turn societal level. As a result, T. Colborn underlines the need for leaders with environmental awareness to be involved in the organizational behavior of the near future, and explains that the most important theme is the cultivation of human resources and organizational design for the next generation that will facilitate decision-making and action in the interests of the global environment.

The qualities required of the leaders of the near future will include decision-making capacities that can prevent action with a negative impact on the environment and can apply lessons from past failures such as corporate pollution and organizational disasters. The global ecosystem is a single integrated bio-system and forms the 'amniotic fluid' for all life. This makes it essential to nurture human resources in the form of sound decision-makers who will have a comprehensive understanding of the global environment. The negative legacy of human cooperation seen in the example of environmental hormones has destabilized the global ecosystem. Remedying this situation will not only require action by the International Organization for Standardization and national governments, but also likely require regulation at the global level by the United Nations and similar organizations. Preventing the spread of environmental hormones is thus a global issue and belongs to the mission of the United Nations and other global organizations.

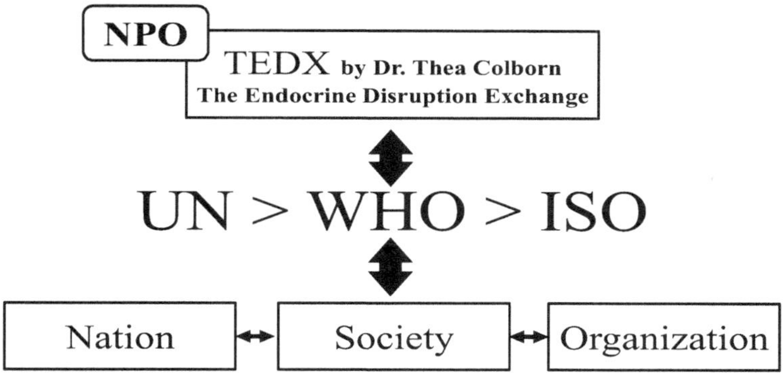

Fig. 3.8 Environmental-hormone regulatory structures. *Note:* Adapted from TEDX URL: http://endocrinedisruption.org/

Societies which have fallen victim to environmental contamination have a safety deficit, because of which the cultivation of sound leaders and decision-makers is an urgent task for the prevention of ecosphere contamination (Fig. 3.8).

The large-scale mass production, mass distribution, and mass consumption based on industrial manufacturing which have emerged in modern organized society and the environmental hormones released by incineration now stand in the way of humankind's survival. To regulate the design of the near future, we need a social code of conduct for global environmental problems. K. J. Arrow commented on 'the moral hazard' created by a society that gives priority to the economy, [14] while M. J. Sandel argued the need for a new morality "from market economy to market society" in step with contemporary trends [15].

Dioxins, which are one member of the group of environmental hormones, are powerful chemicals. Just one drop in the vast space of the Tokyo Dome would apparently be enough to paralyze the nervous system of everyone in it [16]. The formaldehyde contained in adhesives, which is the cause of 'sick-house syndrome', is just one of the many harmful chemical substances surrounding modern humans, which, if used wrongly just once, could have irreversible consequences. However, if their harmful nature is not scientifically proved case by case, it is impossible to stop their manufacture. Even though many people may have found them empirically to be dangerous, if this has not been demonstrated scientifically with numerical data, regulation is impossible under current law. If we compare the present social system to a living organism, we might conclude metaphorically that the parasympathetic nerves are not responding to the sympathetic nerves.

In recent years, the environment-conscious international standards established by the International Organization for Standardization (ISO), which are applied to the production process of actual manufactured products through environmental management and environmental-management auditing (ISO14000 series), [17] have spread from the European Union to become a worldwide standard to which the United States has also subscribed. Going forward, an environment is developing in which it will be impossible for corporations not certified under this international standard to distribute overseas their manufactured products, intermediate products, or services. In the modern world, single-state policy and law have become powerless against environmental problems at a global level, and this will no doubt ensure that the ISO is accepted as a global standard. Environmental management can be seen as fulfilling the role of an international 'business ethic' which transcends the bounds of the state and other systems.

3.2.3 Environmental Management under the ISO 14000 Series

Environmental management is a system aimed at the prevention of global environmental destruction and contamination which clearly delineates the social responsibility of manufacturing enterprises that supply goods and services for the by-products and waste materials they emit by requiring disclosure of information on production systems and organizational control. Its regulation establishes an environment in which the products of enterprises not certified under the system cannot be sold on the international market. And indeed, Japanese industry, and especially the manufacturing industry, is currently making haste to obtain certification under the ISO 14001 environmental-management series. Although environmental management is insufficiently effective as a tool to prevent contamination and destruction, it will probably function as a set of basic preventive rules for the international market. Relatedly, the cultivation of human resources to facilitate sustainable decision-making and action with respect to environmental responsibility is required. At last, the industrial world has begun to recognize the environmental connection and view it as a problem. Environmental management—organizations and the environment presents the ISO vision of the interrelationship between organizations and the environment (Fig. 3.9).

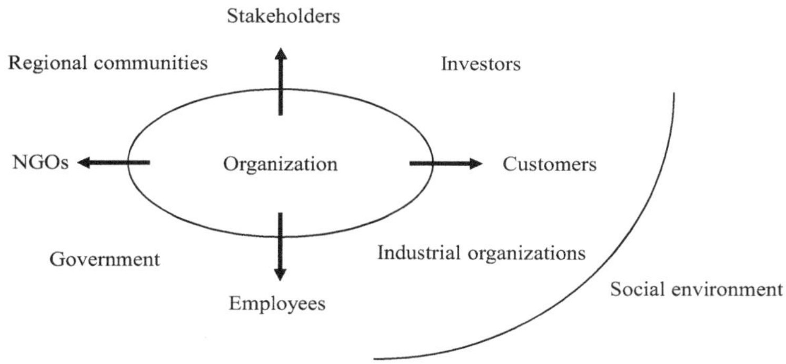

Fig. 3.9 Environmental management—organizations and the environment. *Note:* Charting based on ISO 14001 environmental-management series and A. Gore, Earth in Balance (1992)

Under the current legal system, regulation is impossible unless there is scientific proof in the form of figures and other data on the effect in terms of environmental destruction and contamination. In the cases of the industrial-waste incident at Teshima in Kagawa Prefecture and the dioxin problem at the waste-incineration facilities of the municipalities of Hashimoto and Nose, even though the danger was known, it was not acknowledged publicly in central-government policy or local-government action. Even where actual damage has been reported locally and there is empirical and statistical indication of risk, the government requires data from experiments on guinea pigs. This is the same situation with the cat experiments in the 'Minamata disease' mercury-poisoning scandal involving the Chisso Corporation. In conspicuous contrast to the government attitude is the shared awareness and sympathy expressed at the level of the individual resident or citizen in cases of environmental contamination. Today, there is increasing activism in support of environmental protection by local residents, including through non-governmental and non-profit organizations, and a network for 'environmental ethics' is beginning to form at world-citizen level [18]. One expert emphasizes that there is a shared sensitivity toward environmental protection, and that this network of common sympathies embraces an ethical view of the environment. He attaches importance not to the state or to organizations, but to social awareness based on links between individuals.

This expert is H. R. Maturana, one of the originators of the theory of 'autopoiesis', [19] who gave a lecture entitled 'Ethics as a Human Action' at the 2000 World Congress of Systems Sciences [20]. H. R. Maturana states that the formation of ethical views in society arises from human emotion. In other words, humans and animals form groups, and as part of 'the rules of the herd society', a code of conduct around acts which are bad and acts which are good becomes implicitly shared among group members. All lifeforms that live communally share a common code of conduct for the species with implicit provisions. These implicit rules are important in the survival of the species and constitute what might be called a 'program for the preservation of the species' according to which it is emotion-based behavior that is used to detect and avert danger to ensure self-preservation. It is thus emotion that constitutes the 'operating system' for the survival of humankind, and this cultural network held in common within the social group can be called 'morality or ethics'.

3.3 Organizational Disasters and Environmental Ethics

3.3.1 Organizational Disasters, Accidents, and Pollution

In recent years, with the shift from iron to plastics, synthetic chemical substances which are lightweight, durable, and easily processable have come to occupy an ever-more-prominent place in everyday life. However, the chemical substances

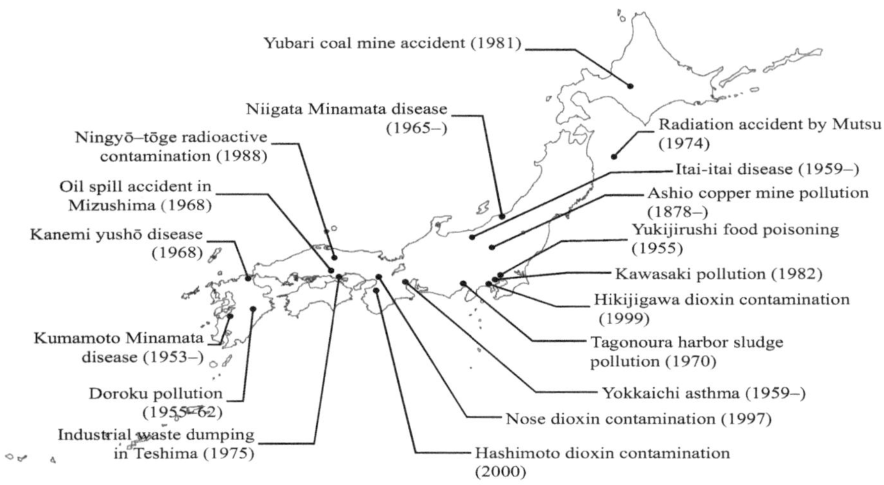

Fig. 3.10 Map of major Japanese pollutions. *Note:* Mapping based on a suggestion by K. Hioki at Kyoto University, 2014

known as plastics, which are mass-produced and mass-distributed for the purpose of mass consumption, not only produce chemical by-products following disposal and mass incineration, but have also been shown to cause contamination to the human body through 'leaching' of harmful substances during their utilization. Already, dioxins and a range of other environmental hormones, either contained in synthetic preservatives, colorants, flavorings, herbicides, PCBs, and agricultural chemicals such as DDT, or released through the incineration of garbage and industrial waste, have spread through many areas of our daily life. The accumulation in the human body of DDT, benzene hexachloride (BHC), and other agricultural chemicals and the finding that PCB contained in breast-milk accumulates in the hair of newborn infants are indicators of the dangers that environmental contamination presents to the next generation.

Figure 3.10 presents a map showing historic pollution incidents in Japan. In the past, corporate-pollution incidents have included the outbreak of the so-called *itai-itai* disease, caused by cadmium poisoning, and the Chisso-Minamata disease, caused by organic mercury. Among cases of pharmaceutical damage, the outbreak of HIV infection caused by tainted blood products, in which hemophilia patients were infected with HIV after receiving 'non-heat-treated' preparations, is still fresh in our memories. The infection was not restricted to hemophiliacs, but spread tragically to their spouses and partners and even their children. There had been earlier cases of pharmaceutical damage, such as when thalidomide and other anti-morning sickness agents caused birth defects, miscarriages, and stillbirths. Why do we continue to repeat the same mistakes when we have already experienced many instances of pharmaceutical damage?

Are these organizational disasters and human-made accidents, of which these pharmaceutical-damage incidents are typical, the inevitable fate of humankind?

Among the organizational accidents and disasters which have come to light in Japan in the last few years are a food-poisoning outbreak in the milk-beverage industry and issues with recycled cow's milk and processed milk; BSE (mad cow disease) and a 'fake beef' scandal, in which the responsibility of the regulatory authority was questioned; the cover-up of medical accidents and mistakes; and a host of other cases too numerous to mention of antisocial activity by organizations. The reason why these organizational disasters attract attention is that activity at the organizational level has a huge effect on the social environment and the bio-environment, sometimes with irreversible consequences. Since organizational disasters are instances of human-induced disaster and human error arising from cooperative activity, there is an issue of both social and historical responsibility. Organizational disasters of this kind present an opportunity to question the morals and ethics not only of central and local government and other public-sector institutions, but also of private-sector enterprises, universities, hospitals, and indeed all organizational entities.

The morality of an organization is an element of the social infrastructure which lies at the foundation of corporations, universities, hospitals, and all other communities, whether in the private or public sector. The conventional social organization maintains itself through pursuit of profit based on economic value and efficiency, and 'organizational rationality' is established as the structural principle. However, maximum profit cannot be realized without survival over time. All organizations living in close contact with society can achieve long-term survival by studying the internal and external environment and contemporary trends and adapting accordingly [21]. C. I. Barnard anticipated this point in his major work, while H. Iino also indicated the importance of 'organizational morality' through the theoretical research [22].

3.3.2 *Creation of Morality by Executive Function*

What in concrete terms is the morality of an organization? In his statements on executive function in Part IV of his major work, C. I. Barnard clarifies the 'role of organizational morality' [23]. Here, he states that the essential nature of business activity is to create new business opportunities by destroying old practices. As examples from modern society, mobile telephones, the Internet, e-mail, portable audio and video players, vehicle navigation systems, and other new goods and services not previously available have been created. By changing our lifestyles, these have changed conventional manners and daily practices. As the goods created by the new businesses alter daily practices, new social conventions that match the times become necessary. One such example is the difference in attitudes between the generations regarding the use of mobile telephones in public buildings and on trains. An important aspect here is that, while businesses with the ability to revolutionize conventional lifestyles create huge markets, they have an inherent potential to lead to social problems. Although the creation of new businesses creates new lifestyles, if matching codes of conduct are not laid down, the innovations will not be accepted permanently. Because business creation destroys

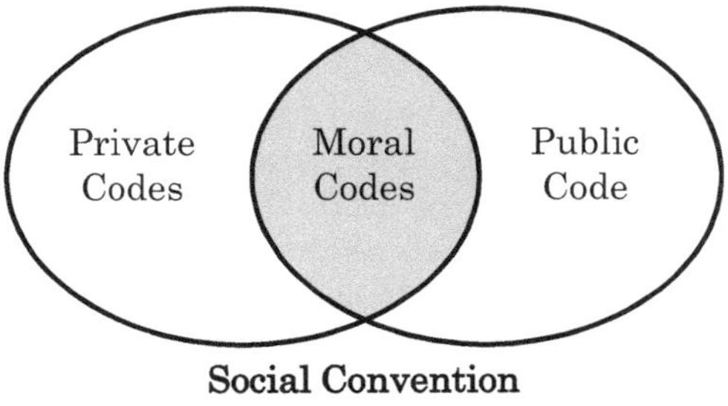

Fig. 3.11 Moral codes bounded by private and public codes. *Note:* 'moral codes' integrate the conflict within each individual between 'private codes and public code'

conventional customs and practices, the role of 'creating new moral codes' is part of the function of executives. That is why C.I. Barnard considered that the creation of morality was a vital part of executive responsibility. In modern society, which is in a state of constant innovation, the morality of the organization, which is intimately connected with social codes of conduct, is today an important management responsibility.

Figure 3.11 illustrates the concept of moral codes composed of 'private codes and a public code' as in Barnard's theory. Organizations that do not fulfill social needs, corporations that do not learn, and other cases where business management disregards the changing times render the respective bodies no longer fit for the modern world and endanger their very survival. Just as lifeforms which do not learn are unable to adapt to environmental change and become extinct, so the social organism represented by the organization cannot survive if it does not learn from the internal and external environment. C. I. Barnard states that "survival is the ultimate measure of the organization" [24]. The morality of the organization, which is a form of social adaptation, is under scrutiny today as an expression of its learning from the internal and external environment.

3.4 Knowledge for Social Survival: Significance of Policy

The modern organized societies of the developed world have in the past extracted fossil fuels, mineral resources, and other raw materials from the earth and processed them for use as thermal and electrical energy in production processes that emit large amounts of waste material and by-products. But now, as part of their environmental strategies, developed countries are actually exporting pollution and industrial waste to developing countries, factory by factory. Manufactured products mass-produced by huge factories for mass distribution and mass consumption are admittedly items which fulfill the everyday needs of clothing, food, and shelter. But when they are disposed of through mass incineration, they cause environmental contamination in the form of dioxins. In modern consumer society, at all stages from pre-production through the production process to post-production, and from pre-consumption through the consumption process to post-consumption, chemical substances not present in the natural world are created, spread through the soil, rivers, lakes, and atmosphere, and cause environmental contamination. As mentioned above, these inflict awful damage on the global ecosphere.

As part of the mixed blessings of the consumer society, chemical compounds which did not exist in the natural ecosystem have been successively created, causing not only contamination of the bio-environment, but also destruction of the social environment. The spread of environmental destruction is brought about by human cooperation centered on organizations. The result of this negative cooperation impacts not only on the natural ecosystem but also on the social environment, something which has begun to endanger the very survival of humankind. Once a substance which the individual cannot create alone has been mass-produced at a corporate factory and mass-distributed, it is quickly incorporated into family life through mass consumption and later disposed of by incineration. The harvest reaped by our consumer society is environmental destruction and contamination. Human knowledge originally used 'organizational cooperation' to reshape the global environment to humanity's convenience. In recent years, however, the water-retention system of mountainsides, lakes, and rivers has been destroyed and the ecosystem thrown out of balance as evidenced by dried-up rivers and landslides. The 'visible hand' of human cooperation has artificially refashioned the natural systems, and this artificial manipulation has altered the global ecosystem. The upshot has been the seemingly irresoluble issue of environmental hormones, which is nothing other than the 'negative dividend of a modern rationality' that prioritizes economic value.

It is intelligence for the sake of survival that defines morality. Living organisms have learned from the environment and adapted. Similarly, the social organism known as the organization has survived by learning from its environment and adapting to the circumstances. The survival of the organization demands a code of conduct for coexistence with the bio-environment and the social environment. In other words, at both the individual and macro levels, the mission of the organism is to learn from the environment and live symbiotically. Both for humans, who are a living organism, and organizations, which are a social organism, there is an essential need for intelligence to control will and action so as to adapt to the environment and survive. In this way, the intelligence of an organism can be seen as a universal principle for the purpose of survival. It is therefore morality, a homeostatic system for the sake of coexistence with the environment, which is the source of knowledge for survival. Just as the senses are important for the survival of the ecosystem, so morality is none other than an algorithm for the survival of the social system.

Conventionally, the approach to organizational behavior which infringed on social convention, such as corporate pollution, was to regulate it retrospectively by law. However, in areas such as global environmental problems and bioengineering, the private sector takes the lead, and legislation and regulation by the public sector have become unable to keep pace with developments. This has created the need for the organization itself to install a mechanism for ethical decision-making. In other words, a structure is needed whereby a moral program can be installed in

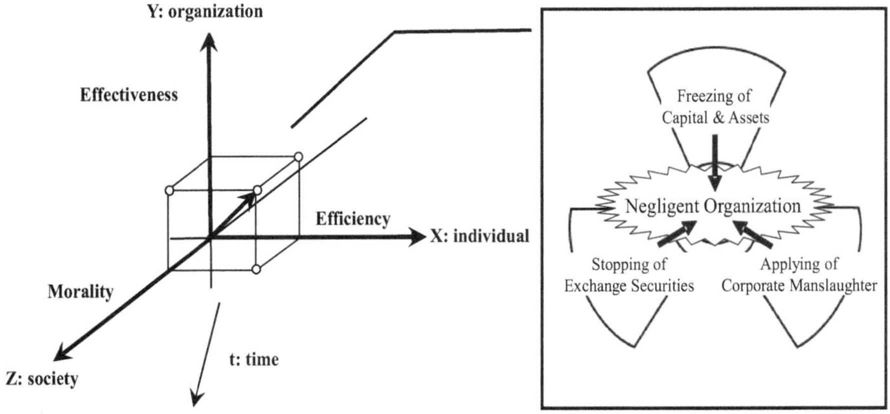

Fig. 3.12 Three-dimensional model of organizational morality. *Note:* Integration based on organizational 'effectiveness', individual 'efficiency', and social 'morality' by C. I. Barnard

the organization itself to control decision-making and action. Going forward, the design of the organizations of the near future should include the cultivation of a moral code that can learn from contemporary trends and cause changes to be reflected in the behavior of its members as part of an executive function fulfilling its social responsibility. Author presents the three-dimensional model of organizational morality shown in Fig. 3.12, in which the two dimensions of organizational effectiveness and individual efficiency are supplemented by a measurement criterion in the third dimension: social morality.

Morality is a concept in opposition to rationality. Modern rationality grew by emphasizing economic efficiency and rejecting human emotion. A society which gave central focus to economic rationality undermined human interrelationships of trust and caused the destruction of morals, ethics, and emotional sensitivity; negated the individual; and gave priority to 'the logic of the organization'. Unfeeling modern society, which excludes the weak and prioritizes material goods, has brought a wealth of mixed blessings. In the interests of an environment-friendly 'sustainable society', [25] the time has come to consider a new model of humanity which will admit a diverse range of value perspectives accessible to all regardless of age or gender. Underpinning the code of conduct that we hold in common is a value system through which we share the emotionality inherent in humanity: in other words, a system of ethics. The 'substance of humanity' spoken of by H. Murata surely refers to the realization of this ethic, which is the foundation common to all humanity [26]. The philosopher T. Watsuji says that 'the embodiment of ethics is policy' [27]. Today, the loss of the ethical view in modern society in all communities, whether industrial, governmental, or Greek academia, calls into question the meaning of *Politike* (policy), which, according to Aristotle, is supposed to realize and embody *Ethica* (ethics).

Environmental hormones could perhaps be described as a kind of test for humankind. The alarm which they sound tells us that consumer behavior prioritizing economic value and focused on material goods has reached its limit and that humanity must shape a new criterion of social welfare. The phenomenon of 'endocrine disruption' caused at the micro level by environmental hormones in the inner body space of living organisms, including humans, is simultaneously mirrored at the macro level of the global ecosystem. Herein lies the reason why questions are being asked about the social responsibility of the decision-makers responsible for mass production and mass distribution and the policy formulators responsible for issuing the relevant manufacturing permits, and why environmental ethics deserves examination as an issue for modern society. To prevent environmental destruction and contamination, there is a growing aspiration for social well-being to be realized in policy formulation and organizational decision-making, not only in the industrial world but also in the worlds of business, politics, and academia. In this context, we look forward to a reconsideration of the meaning of policy, which will hopefully develop as a scientific discipline. Western thought is informed by a basic competitive principle which pits nature against humankind, but in Buddhism the idea of harmony between nature and humanity is at its foundation. With today's urgent focus on global environmental problems, a need has perhaps arisen to move toward an oriental philosophical approach.

References

1. Atsuji, S., "ISO Global Policy toward Environmental Management", *Economic & Political Studies Series No. 135*, The Institute of Economic and Political Studies, 2003, pp. 1–18.
2. Colborn, T., Dumanoski, D. and Myers, J. P., *Our Stolen Future: Are We Threatening Our Fertility, Intelligence, and Survival?*, Spieler, 1996.
3. Colborn, T., Dumanoski, D. and Myers, J. P., Colborn, T., *Ibid.*
4. *See,* http://www.wwf.or.jp/lib/lecture/colborn001028/career.htm.
5. Jacobson, J.L. and Jacobson, S.W., "Intellectual Impairment in Children Exposed to Polychlorinated Biphenyls in Utero", *New England Journal of Medicine*, September 1996.
6. Hauser, P. (eds), *Thyroid Diseases of Infancy and Childhood: Effects on Behavior and Intellectual Development*, American Psychiatric Press, 1999.
7. *See,* http://www.health-station.com/topic92.html.
8. Carson, R., *Silent Spring*, Houghton Mifflin, 1962.
9. Brown, L.R. et al., *State of the World 2000*, W.W. Norton, 2000.
10. Cadbury, D., *The Feminization of Nature*, Penguin, 1997.
11. Brown, L.R., Flavin, C. and Kane, H., *Vital Signs: The Trends that Are Shaping Our Future*, Worldwatch Institute, 1992.
12. Langton, P., *The Black Triangle: Environmental Policy in an International Context*, The Open University, BBC–EPIC, 1999.
13. Brown, L.R. et al., *Saving the Planet: How to Shape an Environmentally Sustainable Global Economy*, World Watch Institute, 1991.
14. Arrow, K.J., *The Limits of Organization*, W.W. Norton, 1974.
15. Sandel, M.J., *What Money Can't Buy: The Moral Limits of Markets*, Farrar, Straus and Giroux, 2012. Sandel, M.J., *Justice: What's the Right Thing to do*, Farrar, Straus and Giroux, 2009.
16. Tsuyumoto, I., *Dioxin*, Natsume, 2000.

17. JACO, *ISO14000 Kankyō management system* [*ISO 14001 Environmental Management System (Bilingual)*], Gyōsei, 2001.
18. Kato, H., *Kankyō to rinri: Shizen to ningen no kyōsei wo motomete* [*Environment and Ethics: Towards the Coexistence of Nature and Humanity*], Yūhikaku, 1998, pp. 1–19.
19. Maturana, H.R., Varela, F.J., *Autopoiesis and Cognition: the Realization of the Living*, D. Reidel Publishing Company, 1980.
20. Maturana, H.R., "Ethics–A Matter of Conservation and Change", *Proceedings of the World Congress of System Sciences 2000*, ISSS, pp. 26, 91.
21. Barnard, C.I., *The Functions of the Executive*, Harvard University Press, 1938, pp. 261–265.
22. Iino, H., "Soshikidōtoku to soshikibunka", *Barnard–Gendaishakai to soshikimondai* [Organizational Ethics and Organizational Culture, *Barnard—Modern Society and Organizational Problems*], Bunshindō, 1986, pp. 286–290.
23. Barnard, C.I., *op. cit.*, pp. 281–284.
24. Barnard, *Ibid.*, pp. 56–61, pp. 251–252.
25. Atsuji, S., "What is the Basis of Complexity?: Cybernetics on the World Congress of the Systems Sciences", *Social and Economic Systems Studies*, The Japan Association for Social and Economic Systems Studies, 2001, No. 20, pp. 149–158.
26. Murata, H., *Kanri no tetsugaku: Zentai to kō, sono hōhō to imi* [*Philosophy of Management: The Whole and the Part, Methods and Meanings*], Bunshindō, 1984.
27. Watsuji, T., *Ningen no manabi toshite no rinrigaku* [*Ethics as Human Learning*], Iwanami, 1934, pp. 46–63.

Part II
Organizational Accidents

In a lecture entitled 'Words that Touch the Heart', Hiroyuki Itsuki states that, in the Meiji Period, there were interpersonal relationships of trust that allowed mutual emotional understanding and that there were sympathy and empathy for sorrow among neighbors. Since the inception of the modern era, there has been a bias toward economic growth targeted at national affluence, and humanity has become diluted in the pursuit of material wealth. As this orientation toward material affluence has increased, human emotionality has been turned into an abstraction under the shadow cast by the prioritization of economic concerns, and 'emotional wealth' has been sacrificed. The human emotional range represents a universal human sensibility and has nurtured mutual understanding and interpersonal relationships of trust. Regardless of location or historic age, to understand human emotion is to understand the sensibility and the values inherent in humans.

The growth-oriented modern society which denies emotion is a ruthless system that prioritizes efficiency, leading to a negation of humanity. It is the 'emotional interconnection' fostered by interpersonal empathy that also leads to the formation of the ethical perspective which is a precondition for human codes of conduct. Sensibility and volition have a fundamental role in human reason, and their interaction with human behavior nurtures human intelligence. Intelligence itself is formed by a process of complementary interaction between reason and sensibility; reason alone is not the source of intelligence. The knowledge created by contemporary science has the power to change the natural environment, but as knowledge itself is essentially created by human intelligence, the creation of knowledge needs to be rooted in a philosophy of coexistence centered on humanity. It is the essence of this intelligence that represents the fundamental wisdom which ensures the continuation of society. A humanity that has lost this wisdom can be seen as risking the destruction of its civilization.

Chapter 4
Crime or Punishment: Brakeless Accidents without Compliance and Governance

This chapter [1] examines a case study of the JR (Japan Railways) West Accident (Photograph 4.1), the worst railway accident [2] in Japanese history. The purpose of this research is to prevent similar accidents by focusing on 'organizational learning disabilities' [3]. Firstly, a review of a summary of the JR accident is carried out. Secondly, the irrational behavior of the driver involved, which originated in the JR West Company's personnel re-education system, the so-called Nikkin Kyōiku (hereafter the 'Nikkin System') is reviewed. Thirdly, the interference with organizational learning bounded by 'structural inertia' is examined, and finally the 'organizational accident' in relation to the learning disability based on a 'brakeless society' is reviewed. This research is concerned with the concept of a 'brakeless organization' without social corporate governance and compliance.

4.1 Brakeless: JR West Railway Accident 2005

Since the advent of a railway-based society, tragic railway accidents have occurred in all time periods and countries, despite great progress being achieved every year in mechanical technology [4]. For instance, 23 people died in an accident in Canada in 1986, 56 in France in 1988, 101 in Germany in 1998, and, more recently, 71 people lost their lives in China in 2008, all due to railway accidents [5]. Figure 4.1 shows a number of accidents that have incurred multiple fatalities since 1980. Although the countries mentioned comparatively advanced technology, these organizational accidents claimed many lives [6].

JR is the common name of Japan Railways, the largest railway conglomerate in Japan, which has a history dating back to the privatization of the Japanese National

---

Reworking of: Atsuji, S. et al., "Organizational Disaster by Implicit Systems Errors: A Case Study of the JR West Accident", *IFSAM*, 2012. *See, for example*, Atsuji, S. et al., *JR Accident 4.25*, Kansai University, 2007.

S. Atsuji, *Unsafety*, Translational Systems Sciences 7,
DOI 10.1007/978-4-431-55924-5_4

Photograph 4.1 The scene of JR accident Photo. No.1, by *permission* of JTSB. *Source:* Japan Transport Safety Board, JR Accident Report 2007. URL: http://jtsb.mlit.go.jp/jtsb/railway/report/RA07-3-1-3.pdf

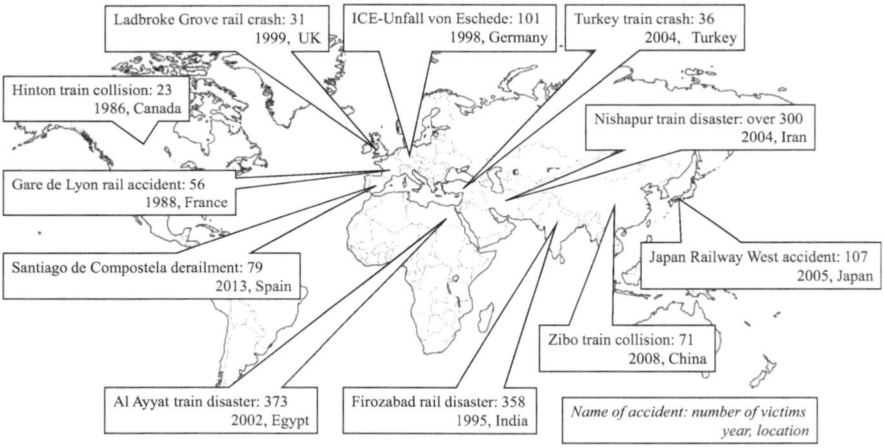

Fig. 4.1 Multiple-fatality railway accidents since 1980. *Note:* Mapping by S. Atsuji and K. Ueda, presented at the Association for the Study of Industrial Management Japan 2010

Railway (JNR) in 1987. On April 25, 2005, a derailment accident occurred on the Fukuchiyama Line of the Japan Railway West Company. One hundred and six passengers and the driver died in this accident, and 562 others were injured. The accident was investigated by the Aircraft and Railway Accident Investigation Commission (ARAIC), whose findings were released as the "Fukuchiyama Line

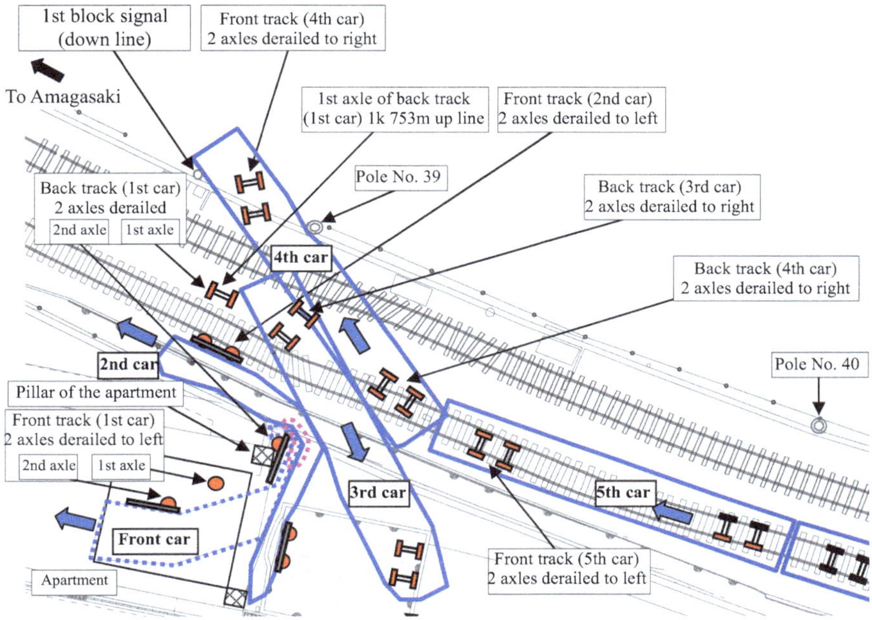

Fig. 4.2 Derailment situation of train, by *permission* of JTSB. *Source:* Japan Transport Safety Board, JR Accident Report 2007. URL: http://jtsb.mlit.go.jp/jtsb/railway/report/RA07-3-1-3.pdf

Derailment Accident Investigation Report" (hereafter, ARAIC's Report) in June 2007 [7]. Figure 4.2 shows the derailment situation of the railcars at the accident site.

Just before the crash, the train overran its intended position at the previous station by approximately 72 m. Because of an adjustment back to correct the location at the station, the train departed from Itami with a delay of 90 sec. It passed through Tsukaguchi, which is the station following Itami on the route to Osaka, with a delay of 60 sec. The train traveled at 116 km/h on this section to make up for the time lost due to the overrun, and then derailed on the curve between Tsukaguchi station and Amagasaki station on the JR Fukuchiyama Line. The excessive speed caused the two front cars to crash into an apartment building after derailment. The upper speed limit at the site was 70 km/h on a curve of 300-m radius. In addition, JR West had a congested railway schedule due to competition with other private railway companies. These situations are the cause of the driver's speeding.

The driver had not performed driving operations for the 40 sec. prior to the accident but had monitored the radio exchange between the conductor and the dispatcher and had made a note of it. The background to the driver's actions was described in ARAIC's report as follows: "There was concern about the Nikkin System (a punitive re-education program for the 'soldiering' on motivation system for JR

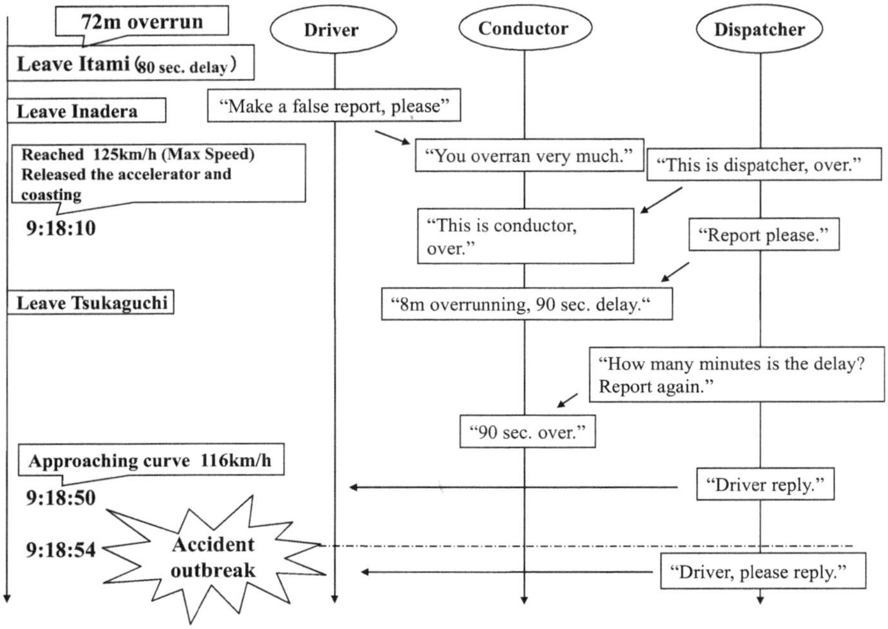

Fig. 4.3 Dialogue from the JR West accident 4.25. *Note:* Drawing by M. Ichimiya and S. Atsuji based on Japan Transport Safety Board, JR Accident Report 2006, pp. 4–16, 34–37. URL: http://jtsb.mlit.go.jp/jtsb/railway/report/RA07-3-1-3.pdf

crew) practiced by JR West, and the harsh measures that were experienced in the past." Figure 4.3 shows the dialogue between the driver, conductor, and control dispatcher immediately before the accident took place (following Fig. 4.14: α).

In this accident, while in actuality the train overran by approximately 72 m at Itami Station, the driver asked the conductor to submit a false report concealing this. The conductor accepted the request from the driver ("please shorten the distance of the overrun") and reported an "8-m overrun and 90 sec. delay" to the control dispatcher of train service management. The control dispatcher made contact with the driver for confirmation. The driver was in a dangerous situation because the reported 8-m overrun is inconsistent with a delay of 90 sec., as became clear from the train service recorder immediately after the train crashed. ARAIC's report noted that the driver's dangerous driving was caused by fear of the Nikkin System, which he had experienced in the past. Figure 4.4 shows the handwritten notes which the driver was taking until the accident took place. The driver took the notes for his self-defense during the 40 sec. when he was not driving, when the train was approaching the curve at 120 km/h, way over the maximum speed limit of 70 km/h on stretches with a radius of curvature of 300 m or less (Photograph 4.2).

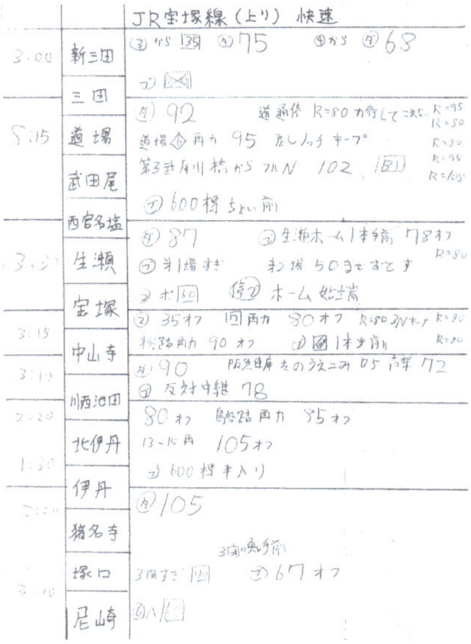

Fig. 4.4 Driver's original memo *Source:* Japan Transport Safety Board, JR Accident Report 2006

Photograph 4.2 The scene of JR accident Photo. No.2, by *permission* of JTSB. *Source:* Japan Transport Safety Board, JR Accident Report 2007. URL: http://jtsb.mlit.go.jp/jtsb/railway/report/RA07-3-1-3.pdf

4.2 Structural Inertia by Misgovernance

The Nikkin System is the re-educational program carried out for the purpose of preventing accidents and incidents, but part of this system consists of punitive measures. The program is performed from 9:00 to 17:45 in the office work room of each train division. This room is a space for office workers and administrators, and those subject to the Nikkin System sit in the position labeled 'driver' in Fig. 4.5. [8]. These personnel are required to work all day on a report under the supervision of an administrator or office personnel. Members who have been subjected to the Nikkin System say "My exposure to the other members made me feel uncomfortable" inside the 'strange world of organizations' [9].

The Nikkin System mainly consists of report writing, and also includes a test that measures the driver's basic knowledge. However, the educator in charge determines the actual work content in the Nikkin System, and questionable chores of a punitive character are also included, such as longhand 'copying of work rules' and 'weeding of train tracks or flower beds', as reported after this accident. In addition, anyone who undergoes the Nikkin System may have his or her salary reduced [10]. Such a punitive system is an example of the type of education method that former Japanese companies and the Japanese armed forces often adopted. One problem associated with this educational method is that it depends excessively on personal spiritual strength and concentration without investigating the cause of the failure within 'unsafe acts' [11].

Fig. 4.5 The situation of Nikkin System at JR West *Source:* H. Suzuki et al., The Mortal Sin of JR West, 2007, p. 67

Table 4.1 Suicides of JR West crew members (from 2000 to March 2005)

Date of suicide	Method	Date of suicide	Method
Mar 21, 2000	Hanging	Apr 21, 2003	Jumping in front of JR train
Summer, 2000	Hanging	Apr 24, 2003	Jumping from JR building
Oct 24, 2000	Jumping in front of train	Jun 23, 2003	Hanging
Jan 10, 2001	Hanging	Jul 20, 2003	Jumping in front of JR train
Jan 12, 2001	Entering the water	Sep 1, 2003	Jumping in front of JR train
Feb 8, 2001	Jumping in front of train	Sep 23, 2003	Hanging
Apr 24, 2001	Jumping in front of train	Jan 31, 2004	Hanging
Sep 6, 2001	Hanging	Oct, 2004	Jumping in front of train
Oct 14, 2001	Hanging	Mar 13, 2005	Suffocation by carbon monoxide poisoning

The above information reveals that those who demonstrated their despair and hatred toward the company by jumping from JR buildings or in front of JR trains they had loved did so as a last resistance
Source: H. Suzuki et al. *The Mortal Sin of JR West*, 2007, p. 164

Table 4.1 shows the number of suicides that have occurred at the JR West Company [12]. From 2000 to 2005, 18 employees committed suicide, and, on average, 4 people take their own lives each year. There are six railway companies in the JR Group, each operating in a separate region: JR Hokkaido, JR East, JR Central, JR West, JR Shikoku, and JR Kyushu. No data exist regarding the number of employees overall who have killed themselves, and only JR West has been brought to public attention. Although it cannot be concluded that the direct cause of these suicides is the Nikkin System, there is the possibility that problems exist under JR West's management (following Fig. 4.14: γ).

The driver involved in the accident described above had experienced the Nikkin System three times, for a total of 18 days. In addition, the driver occasionally complained to his friend that "I must write text all day long and need permission even to go to the toilet," as described in ARAIC's report. Following the accident, on June 1, 2005, a questionnaire was distributed to 3,096 drivers by the West Japan Railway Union, and 2,676 responded. Over 25 % of the respondents answered that JR employees felt dissatisfaction. As stated above, the Nikkin System was a personnel-management system. The purpose and result of this educational method diverged, and, in general, the managers of JR West did not engage in 'double-loop' learning [13].

In the case of this accident, the driver did not operate normally and tried to protect himself from the Nikkin System: that is, the 'un-learning' processes of the JR West organization.

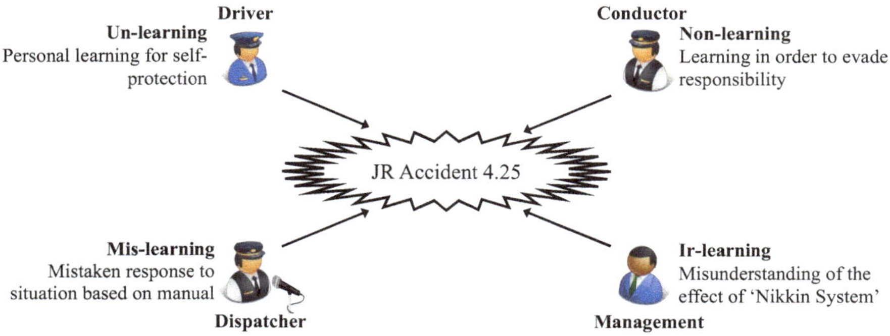

Fig. 4.6 Four learning disabilities by the structural inertia in organization. *Note:* Illustrating based on 'Learning Disability' by D. A. Garvin and 'Organizational Inertia or Culture' by C. Argyris. *See also* 'W-loop Organizational Learning' by J. G. March, R. M. Cyert, and J. P. Olsen

First, the driver requested the conductor to make a false report. Further, the driver made notes in an act of self-protection to avoid having to undergo the Nikkin System in spite of the actual driving operation. The driver's behavior was a personal form of learning for his own self-protection: that is, it was *un-learning*, which means he could learn but refused to do so. Second, the conductor did not use the emergency brake, and worse, he did not know how to use it. In this case, learning did not materialize: that is, the situation involved *non-learning*. Thirdly, the dispatcher made contact with the driver for fact-checking despite the existence of an ongoing dangerous situation. The behavior of the dispatcher followed the manual. However, this action was a mistake resulting from a lack of circumstantial judgment: that is, *mis-learning*. Finally, management's misunderstanding of the effect of the Nikkin System is also involved because they ignored the feedback from company personnel and put profits above safety in their management policy. This decision-making process caused negative effects in learning: that is, irrational learning, or *ir-learning* (Fig. 4.6).

The Nikkin System, as stated above, is a re-education system used for personnel management and is designed to prevent accidents caused by human error against a background of system error [14]. However, this system led to human error. JR West managed employees using the psychological pressure provided by the Nikkin System and by attaching importance to manual labor without feedback. Ultimately, 'multifaceted learning disability' occurred at unit organization levels [15]. The cause of these learning disabilities is the lack of communication between members of the organization, which forbade questions regarding organizational policy and objectives and concealed facts. These flawed traditions and customs led to structural inertia in social organization. That is, the organizational climate and culture reduced the mental horizons of its members, who were unable to think of anything except their own self-protection.

Structural inertia in an organization refers to the internal practices, conventions, and similar aspects of an existing organization which are part of the culture that has become established within its system. In the case of JR West, an irregular form of train crew staff education had been practiced over successive years in the form of the Nikkin System program, which was characterized by punitive sanctions (enclosure) and public bullying (e.g., track-side weeding), indicating the presence of unreasonable operational practices within the organization's systems. In JR West alone, which was just one of the seven companies of the JR Group, 18 employees committed suicide over a 5-year period, leaving behind notes that spoke of suicidal anguish. In this catastrophic railway accident, the human error on the part of the train crew was committed against a background of structural failings in the organization. Among these were (1) the Nikkin System program, which inhibited driver action; (2) failure to install the ATS-P automatic train stop device; and (3) a management approach which prioritized profit, for instance allocating resources to commercial facilities in preference to the safe running of the railway. These points were also seen as problematic, and the possible involvement of organizational system error was considered by the public prosecutor's investigation.

In summary, the disaster cannot be explained simply in terms of a national preoccupation with punctuality; the abnormal organizational culture of Japan's corporate system was also a latent element which encouraged the establishment of a structural inertia that was chronic rather than transient, indicating the deep roots of the problem. The structural inertia of this type of corporate organization surely reflects a structural pathology of 'brakelessness' that affects not only corporate organizations, but Japanese society itself, as indicated by the world's highest incidence of death and suicide from overwork, of which there are more than 30,000 cases a year.

4.3 Lost Compliance by Administrative Limitations

According to Reason [16], 'security holes'—weaknesses and gaps in safety—always exist somewhere even if precautionary safety measures are taken. However, the implementation of repeated precautions can be expected to solve this problem. Unfortunately, accidents can still happen because the holes in safety measures can occur in any location or even move and spread [17]. While further improvements in technology may arise, the problem of railway accidents cannot be solved on a technological basis alone but must also be addressed from an organizational perspective that includes decision-making, personnel education, and policy-making.

Several years after the accident, it became clear that JR West had required that the accident report be concealed from the investigating officials. JR West intervened actively in three ways, as 'representatives' of the public, within the investigating body itself, and among the supervisory authorities. As respective examples, JR West approached a speaker at a public meeting; arranged beforehand to tell the same story to the police; and then demanded from the investigators a change in a report about "a delay in the deployment of the ATS [Automatic Train Stop]."

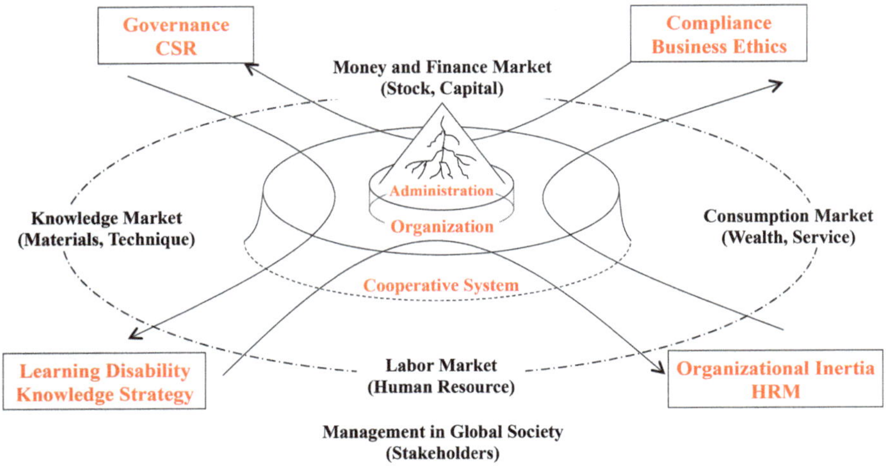

Fig. 4.7 Administrative limitations (management, organization, administration, and society). *Note:* Diagram based on Y. Yamamoto and H. Iino adapted from The Functions of the Executive by C. I. Barnard

According to ARAIC's report [18], the cause of the accident was 'human error,' i.e., the accident was attributed to a delay in braking. In addition, these reports presumed that this accident could have been avoided by the Automatic Train Stop (ATS) device. The ATS is a form of safety equipment designed to stop or decelerate trains to ensure safe operation [19]. There are several types of ATS devices, and JR West has sometimes used a newer type (ATS-P) in place of the old type (ATS-SW) since 1990. In fact, the old type of ATS is incapable of stopping trains that are speeding. On the day of the accident, April 25, 2005, ground ATS equipment was not installed in this railroad area—not even the old type of ATS. As the curvilinear speed was excessive, the accident was labeled the result of human error. Moreover, this case involves double standards regarding installation of the ATS device, as both the SW and P types were used by the supervisory authorities as the result of governmental policy. The government authorized use of the old type of ATS only for JR, whereas the major private railroad companies were required to install the newer type of ATS.

Public and business administration management in the institutional organization, as illustrated above (Fig. 4.7), is conducted through the medium of the organization as a cooperative system based on the complex whole constituted by the managerial resources of people, goods, money, and knowledge [20]. It displays presence, structure, and process through the three-layer model of cooperative system, organization, and administration, forming a 'trinity of business', corporation, and management [21]. *Human cooperation* thus appears to be realized through the action of cooperative system as medium, organization as reactant, and administration as catalyst. However, the human-made disasters including organizational accidents and corporate pollution of recent years indicate that there are also many cases of deterioration into dysfunction and systems pathology. Given the global change in the social environment and the ecosystem, it is essential for governance and compliance [22] to operate as social functions (Fig. 4.7, above).

4.3.1 Lost Compliance: Fuzzy Policy

ATS Double Standard (Ambiguous Railway Policy)

It has been suggested that part of the reason for the delay in installing the ATS device was that the Japanese government's policy on the installation of the new ATS-P operated a double standard that discriminated between the former state-owned Japanese National Railway (JNR) and private railway companies. Under government railway policy, installation of the new ATS-P with speed control was compulsory for major private railway companies, while the former state railway company alone was allowed to use the older SW version of the ATS. Regarding ATS installation, there was thus a double standard which permitted the former state railway company (nowadays the JR Group) to avoid installing the new ATS-P.

4.3.2 Misgovernance: Failure Management

Imbalance in Staff Recruitment (Management Failure Through Staff Cuts)

The former JNR which was the predecessor of the JR Group had a staff of 477,000, but following the privatization and breakup of 1987, staff cuts came to be seen as the essence of effective management, and the new JR companies decided to take on only around 200,000 staff from the old state railway [23] (Fig. 4.8).

For a period of some 10 years starting even before the restructuring of the state railway company, JR West undertook no recruitment of drivers, conductors, or other train crew staff, so that by the time of the accident there had been a great drop in the number of staff in their thirties. Specifically, of JR West's total of 4,233 drivers, only 19 were in their thirties. It has been suggested that the intergenerational transmission of driving skills and knowledge was impaired as a result. The lowering of skills resulting from this imbalanced staff structure was seen as an issue in Human Resource Management [24] (Fig. 4.9).

Union Members' Questionnaire Evaluation

The Nikkin System operated with the supposed aim of preventing accident recurrence, but drivers and conductors at the frontline took a dubious view of the

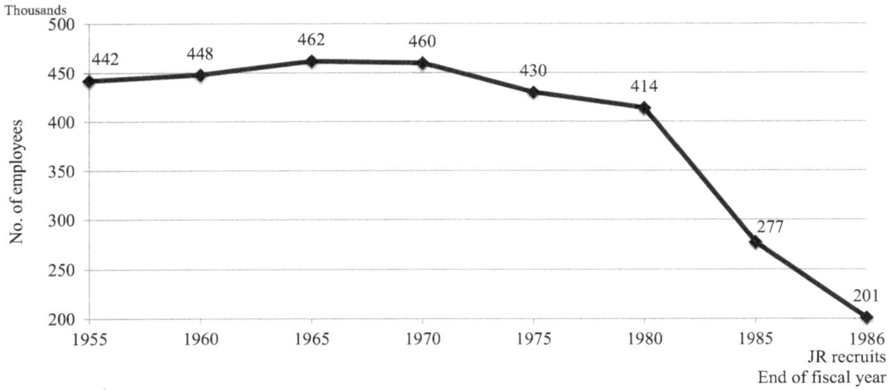

Fig. 4.8 Change in Japanese National Railway staff levels *Source:* Ministry of Land, Infrastructure, Transport and Tourism, Railway Bureau Report

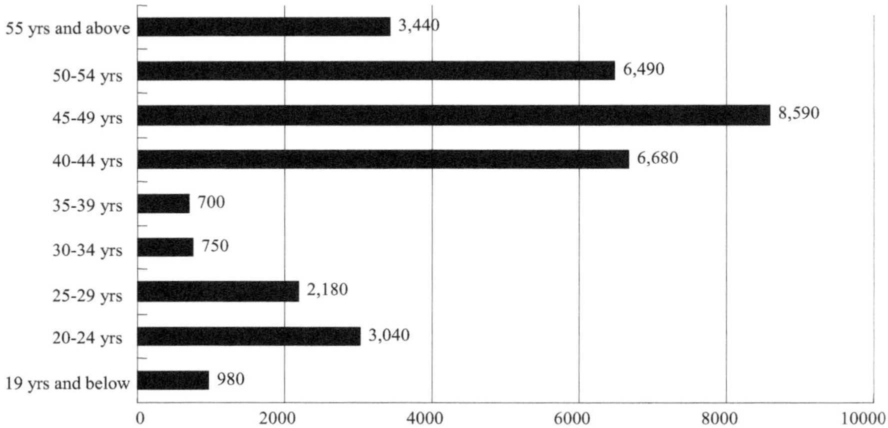

Fig. 4.9 JR West staff structure by age (2004). *Note:* 'Knowledge transmission' of 30's generation gaps based on a discussion with G. Fink at IFSAM Berlin 2006 *Source:* Abe [24]

educational content of the program. According to a questionnaire, only around one quarter responded that they found the content of the Nikkin System program 'appropriate', indicating dissatisfaction with the content of re-education. The questionnaire in Fig. 4.10 was conducted by the labor union of the West Japan Railway Company, and was distributed on June 1, 2005, to 3,096 train crew staff, with 2,676 responses received, a response rate of 86.4 %. In the multiple-choice answers regarding the content of the Nikkin System program, one of the most frequent responses to the question of what disciplinary measures staff had faced was that they had been made to write a statement recognizing their error in the presence of the other train crew staff. Regarding verbal criticisms received, among the most

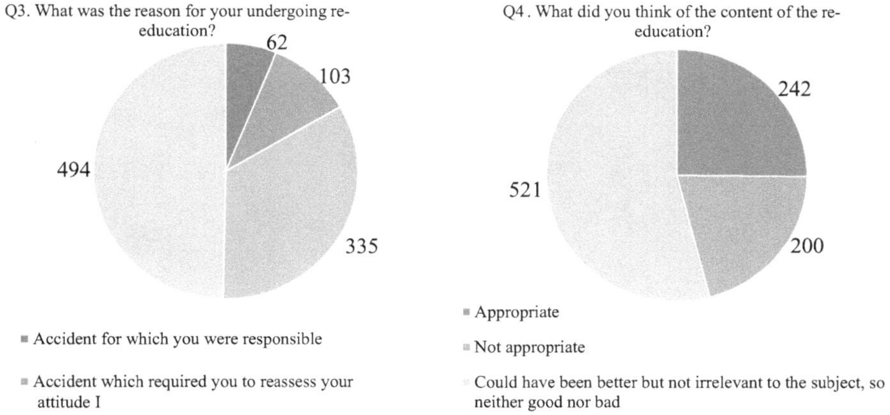

Fig. 4.10 JR West labor union's urgent questionnaire concerning the Nikkin System. *Note:* Calculated using data from West Japan Railway Company labor union (June 1, 2005)

common answers were being told to resign as driver or being told that they would be made to resign the next time they made a mistake [25] (Fig. 4.10).

Because the content of the Nikkin System program was 'punitive', many staff members were apparently worried about being sent on the program. This constant threat was a psychological burden to train crew staff, and the possibility cannot be ruled out that it drove the kind of behavior displayed by the driver and conductor in the accident (false reporting, listening in on radio exchanges, taking notes). On the other hand, the management team's appraisal of the Nikkin System after the accident, expressed in the statement "pressure is necessary, it is a useful thing", indicates an understanding at odds with that of the frontline workers.

Allocation of Management Resources

An investigation of the reason for the delayed introduction of the ATS-P device on the Fukuchiyama Line shows that the equipment investment plan was decided on the basis of the number of passengers and service frequency alone. As the number of trains was fewer than on the Hanwa Line, the Osaka Loop Line, the Kyoto Line or other major lines, the installation was apparently postponed. Since fiscal 1989, JR West has introduced ATS-P successively on major lines in the Osaka region. The installation of ATS-P on the JR Fukuchiyama Line (Takarazuka Line) where the accident took place began in 2004 and ATS-P operation began in June 2005, but at the time of the accident installation had not yet taken place [26] (Fig. 4.11).

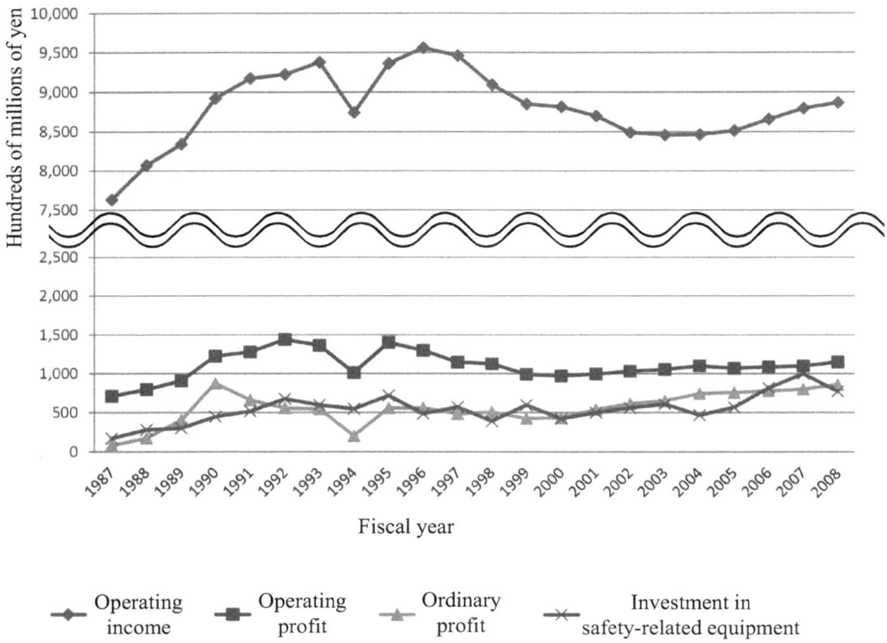

Fig. 4.11 JR West management performance and investment in safety equipment. *Note:* Calculation by K. Ueda, R. Fujimoto, and S. Atsuji based on JR West in Figures 2008, JR West Railway Company, p. 28

4.3.3 Non-CSR: Antisocial

Compensation Negotiations for the Bereaved Families

JR West has taken the opportunity at public meetings and elsewhere to apologize to the bereaved families and injured victims, but has been slow to provide compensation and restitution. As late as 2007, more than 70 % of the bereaved families are not participating in the compensation negotiations announced in that year. Even in 2014, 9 years on from the accident, the *Kobe Shimbun* newspaper of April 21 carried on page four a survey of the bereaved families in which 30 % claimed they were not participating in the compensation negotiations while 30 % said they had already reached agreement, indicating that progress in resolving restitution and compensation issues had been slow.

Coordination of Police Statements Among Employees

JR West also attempted to obstruct the police investigation internally. A study meeting to prepare employees for police interviews was held to rehearse the questions envisaged from the Hyogo Prefectural Police. Statements from JR West executives repeated almost identical wording in the same order in phrases such as "words of apology to the bereaved families and injured passengers", "appropriate safety measures had been taken", and "the danger of the curve could not have been foreseen". According to the investigating police officers, the responses were 'cookie cutter' statements. Moreover, in searches at the homes of JR West staff, the Kobe District Public Prosecutor's Office seized a distributed document entitled "Questions likely to be asked by the police". This series of actions on the part of JR West, which hindered the public prosecutor and police investigations of the JR accident, constituted antisocial action, infringed against compliance, and raised questions of corporate social responsibility.

Payments to Individuals Who Spoke at a Public Meeting

JR West provided retired employees of the state railway company, JNR, who spoke at a public meeting about the accident with documentation prepared by JR, made a recruitment appeal for people willing to speak at the meeting, and made payments to them of 100,000 yen. In early January 2007, the then chairperson of the JR West Safety Advisory Panel leaked to JR the draft version of a report to be presented to the government. Meanwhile, on December 27, 2006, JR West requested the retired JNR employees to make public statements based on their view of the railway accident, in response to which they issued the convenient opinion that "ATS should not be installed at the curve". After the public meeting, according to Mr. X, CEO of JR West at the time of the hearing, the retired JNR employees received a cash payment of US$1,000 from JR West for the preparation of their statements [27] (Fig. 4.12).

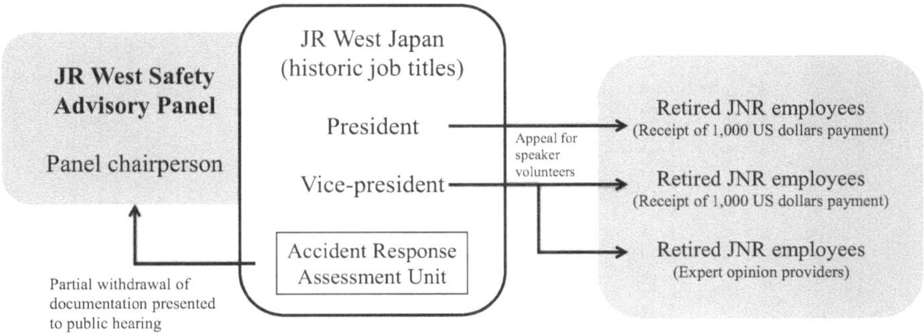

Fig. 4.12 JR West influence on individuals making public statements. *Note:* Presented by K. Ueda, R. Fujimoto, and S. Atsuji at the Association for the Study of Industrial Management Japan, 2009. Drawing based on Nihon Keizai Shimbun, Asahi Shimbun, Mainichi Shimbun, Yomiuri Shimbun, and Kobe Shimbun newspapers

4.4 Guilt and Punishment of Brakeless Organizations

A human-related factor is involved in the problems with the ATS: namely, managerial decisions. There was a large difference in safety measures between JR West and JR East, thus within the same industry. JR East allocated a budget of 25 billion yen to safety measures, but JR West allocated only 3.5 billion yen. On the other hand, JR West greatly reduced its deficit after having switched from being a public enterprise to a private enterprise. It is thought that the managers of JR West might lack social responsibility in the decision-making process regarding the distribution of managerial resources. This amounts to an organizational cause of the accident, the JR accident originating, as stated above, not only from technological factors caused by lack of ATS equipment due to delays in installation, but also from psychological factors present in the organizational personnel system that even extended to bribery. These are behaviors that completely contradicted managers' apologies to victims. As Fig. 4.13 illustrates, JR West has engaged in illegal behavior with 'lost compliance'.

According to Berle et al. [28], governance is established based on the propriety of managerial power. Governance functions only if the managerial power rests on the premises of social propriety and neutrality. However, the illegal acts of JR West lie far from a sense of social ethics. The propriety on which Berle insists cannot be

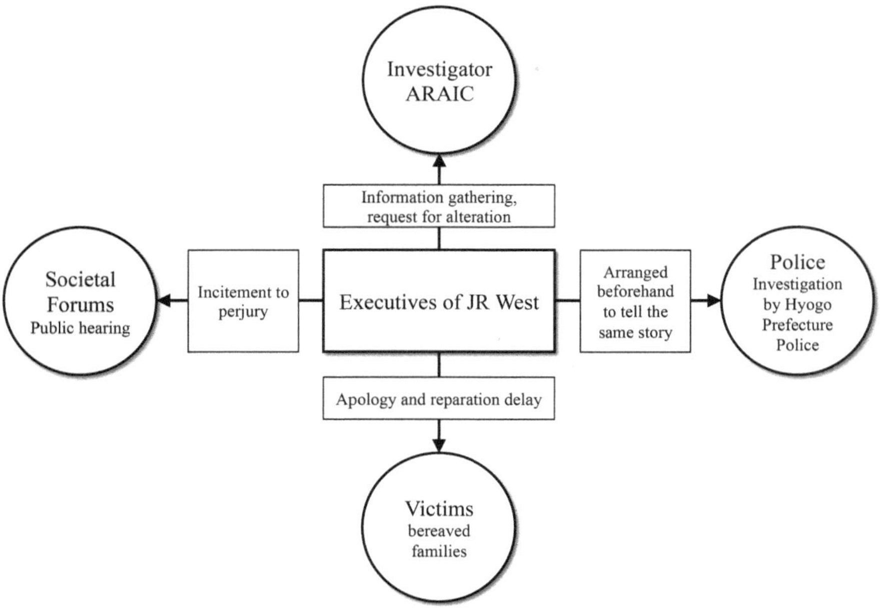

Fig. 4.13 Distrust: lost compliance and governance by mis-leadership. *Note:* Charting based on a discussion with F. Isada and K. Ueda and inspired by G. B. Graen at IFSAM Limerick 2012

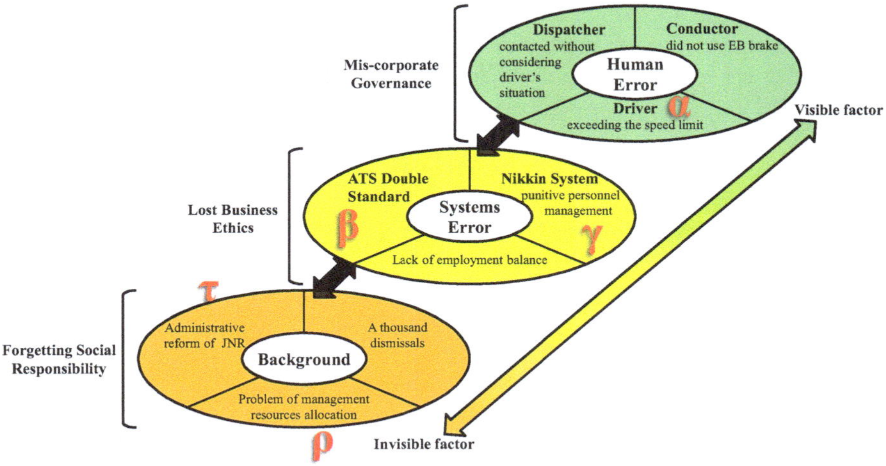

Fig. 4.14 Invisible mutuality of the JR accident. *Note:* Diagram based on discussions with K. Ueda, R. Fujimoto, M. Fujii, and M. Ichimiya in Atsuji Seminar

identified in this case. A series of organizational injustices carried out by JR West represents negative governance based on the wrongful use of power.

Figure 4.14 shows the complicated factors in the JR accident. A summary of this accident indicates problems at three levels. Firstly, as stated above, employees had learning disabilities at each level, representing the human errors that caused this accident. In this respect, JR West's management could be described as 'misgovernance'. Secondly, the problems with ATS and the Nikkin System were the results of 'lost business ethics'. Human errors should be accounted for by the system, but in this case, the system itself embodied errors. Finally, some organizational factors existed: for instance, privatization and 'a thousand dismissals' (*see below*). In fact, the main problems underlying this accident are the culture and climate that were in place during the long history of the organization [29]. This aspect can be termed 'forgetting social responsibility'.

There are many factors at work in the historical background to the JR West accident on 4.25. For instance, the administration of JNR was responsible for the reconstruction of the organization that predated JNR—the so-called Kokutetsu. The Kokutetsu was protected by the Japanese government and had an entrenched 'structural inertia'. Furthermore, JNR's organizational members are 'yes-men': that is, employees who profit anytime they agree with management decisions. On the other hand, the negative aspect was the learning disability inherent in their organizational change. When reforming JNR, these agents had to undertake a thousand dismissals. This meant the laying-off of a thousand employees of JNR, with the decision-making being top-down. At the present time, the matter is pending in court. Many problems in the historical background have thus gradually begun to surface in connection with allocation of HRM (Human Resource Management) [30]. That is the 'invisible factor' of the historical aspects.

In light of the abovementioned JR accident, it is believed that the degree of danger involved in organizational accidents should be established. Specifically, a formula for organizational accidents based on empirical data from the JR accident is suggested.

4.5 Formula for Organizational Accidents

$$\varepsilon = \alpha \times \beta \times \gamma / (\rho + \tau)$$

α: excessive speed
β: ATS problems
γ: Nikkin System
ρ: safety management
τ: railway policy

The formula for organizational accidents relies on five variables, for which empirical data are readily available: the human factor (in the case of JR, the driver's speeding and system of contact (α)), the physical factor (the 'double standards' concerning ATS (β)), unexpected factors (the overrun and delay, and fear of the Nikkin System (γ)), the systemic fatigue combined with the a reliability of the 'safety management' (ρ) and the assessment of the 'railway policy' (τ). This formula includes not only visible factors such as human error but also invisible factors such as the railway policy and business management situations, including a rating reliability by the leading rating agencies (Moody's and Standard & Poor's).

The trial of the ex-president of JR West was completed on January 11, 2012, and the judgment was 'not guilty'. He recreated the crash site as it originally was and shortened the curve from 600 m to 304 m in 1996. He was also the person responsible for ATS equipment. Additionally, in court, there were displays of organizational antisocial behavior by JR West. These unsafe behaviors were intended to protect the corporation and individuals themselves, resulting in organizational inertia. Such organizational inertia demands that 'compliance' and 'corporate governance' be rejected.

Author show the new administrative measures. If the organization causes some problems or is involved in a scandal, it is necessary to regulate it in an external environment (Fig. 4.15). For instance, the supervisory authorities can forcibly freeze the capital and assets of a company when the organization practices misgovernance, and can stop the exchange of securities when there is a loss of business ethics. Furthermore, stakeholders can audit the market for signs of forgetting social responsibility. In order to regulate in such ways an organization that has negative inertia, it is essential to give the supervisory authorities the executive ability to regulate organization.

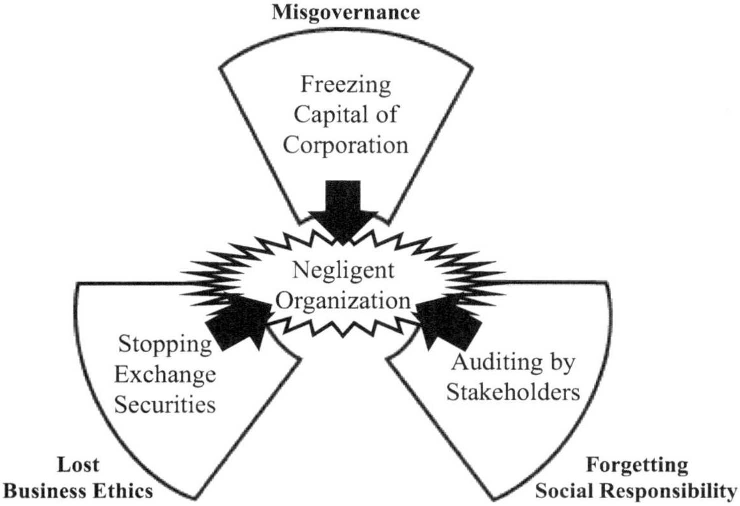

Fig. 4.15 Regulatory system against negligent organizations. *Note:* Charting "Regulation for the negligent organizations based on Organizational Accident" by J. Reason, 'Disaster Management' by S. Bennett, and "Corporate Manslaughter" in UK Law

The longer the history of an organization, the more strongly organizational inertia can develop, and this inertia cannot be changed easily. When similar accidents occur, individuals can evade individual responsibility because these accidents are not caused by individuals but are the result of organizational behavior. Great importance is attached to the creation of new laws that can judge organizational behavior. However, that is the role of government. JR West should reconsider the relationships of various stakeholders based on their social responsibility with compliance and governance as a public-service organization by making use of a review of the failures involved, and they should further undertake a radical rebuilding of the decision-making process of the organization.

Railway accidents cannot be solved only from a technological viewpoint but also must be addressed from an organizational perspective, using the case study of the JR West accident as the worst-case 'organizational disaster'. The purpose of this chapter is prevention of similar accidents, by focusing on organizational learning disability. To review, firstly, a summary of the JR accident; secondly, the irrationality of organizational behavior, originating in the re-educational system of the JR West Nikkin System by their lost compliance; and finally, the interference of organizational inertia in members' learning disabilities by their misgovernance have been covered.

References

1. Atsuji et al. "Organizational Disaster by Implicit Systems Errors: A Case Study of the JR West Accident", *IFSAM*, 2012. *See for example,* Atsuji et al., *JR Accident 4.25*, Kansai University, 2007.
2. ARAIC (2006), The Aircraft and Railway Accident Investigation Commission, *Jijitsu ni kansuru hōkokusho no an ikenchōshukai* [*Draft Report Concerning Fukuchiyama Line Derailment Accident*], 2006.
3. Garvin, D.A., *Learning in Action: A Guide to Putting the Learning Organization to Work*, Harvard Business School, 2000.
4. Reason, J., *Review. I. Overview. II. Theory,* Derby: British Railways Board, 1993.
5. Nikkei Telecom 21, Newspaper article databases (http://edu21.nikkeimm.co.jp) (*Accessed* on November 13, 2006).
6. Hidden, A., *Investigation into the Clapham Junction Railway Accident*, Department of Transport. London: HMSO, 1989. *See also,* Free, R. 1994. *The Role of Procedural Violations in Railway Accidents,* PhD Thesis. University of Manchester.
7. ARAIC (2007), The Aircraft and Railway Accident Investigation Commission, *Tetsudō jiko chōsa hōkokusho Nishinihon ryokaku tetsudō kabushikigaisha Fukuchiyama~Amagasakikan resshadassenjiko* [*Fukuchiyama Line Derailment Accident Investigation Report*], 2007.
8. Suzuki, H. and Yamaguchi, T., *JR nishinihon no taizai: Hattori untenshijisatsujiken to Amagasakidassenjiko* [*The Mortal Sin of JR West*], Satsuki Publishing, 2007, p. 67.
9. Mintzberg, H., *Mintzberg on Management: Inside Our Strange World of Organizations*, New York: The Free Press, 1989.
10. Reason, J., Parker, D. and Free, R. 1994. *Bending the Rules: The Varieties, Origins and Management of Safety Violations,* Leiden: Faculty of Social Sciences, University of Leiden.
11. Reason, J., *The Human Contribution: Unsafe Acts, Accidents, and Heroic Recoveries,* Ashgate Publishing, 2008.
12. Suzuki, H. and Yamaguchi, T., *op. cit.,* p.164.
13. March, J.G. and Simon, H.A., *Organizations*, Wiley & Sons, 1958. *See*, March, J.G. and Olsen, J.P., *Ambiguity and Choice in Organizations,* Universitetsforlaget, 1976 and 1979. *See also,* Cyert, R.M. and March, J.G., *A Behavioral Theory of the Firm*, Prentice-Hall, 1963.
14. Reason, J., *Human Error*, New York: Cambridge University Press, 1990.
15. Argyris, C. and Schön, D., *Organizational Learning: A Theory of Action Perspective*, Addison-Wesley, 1978.
16. Reason, J., *Managing the Risks of Organizational Accidents*, Ashgate Publishing, 1997, pp. 11–13.
17. Fennell, D., *Investigation into the King's Cross Underground Fire*, Department of Transport, London: HMSO, 1988.
18. ARAIC (2006), *op. cit., See also*, ARAIC (2007), *op. cit.*
19. Meister, D., "Human error in man-machine systems", in Brown, S. and Martin, J. (eds), *Human Aspects of Man-made Systems,* Milton Keynes: The Open University Press, 1977.
20. Kataoka, N., "Keiei Kouzou Bunseki–Jigyou, Kigyou, Keiei", Yamamoto and Katou, *Keieigaku Genron (Principle of Management)*, Bunshindo, 1982.
21. Hioki, K., *Keieigaku Genri (Management Fundamentals)*, Economist, 2000.
22. Ota, S., *Kigyou to Seisaku (Business and Policy)*, Minerva, 2013.
23. Ministry of Land, Infrastructure, Transport and Tourism, Railway Bureau website.
24. Abe, S. (eds), *Public Transport is Dangerous: Deregulation and Work Overload*, Iwanami Shoten, 2005, p. 11.
25. West Japan Railway Company labor union, June 1, 2005.
26. JR West, *JR West in Figures 2008*, JR West, p. 28.
27. *Nihon Keizai Shimbun*, *Asahi Shimbun*, *Mainichi Shimbun*, *Yomiuri Shimbun*, and *Kobe Shimbun* newspapers.

28. Berle, A.A. and Means, G.C., *The Modern Corporation and Private Property*, Macmillan, 1932, pp. 356–357.
29. Weick, K.E., *The Social Psychology of Organizing, second edition*, The McGraw-Hill Companies, Inc., 1979.
30. Morley, M., Heraty N. and Collings, D. (eds.), *International Human Resource Management and International Assignments*, London: Palgrave Macmillan, 2006.

Chapter 5
Lost Trust: Socio-biological Hazard—From AIDS Pandemic to Viral Outbreaks

Iatrogenic HIV infection refers here to cases of infection with the human immunodeficiency virus (HIV) caused by private and public administration of blood products [1]. Following the discovery of acquired immunodeficiency syndrome (AIDS) in 1981, numerous warnings were issued by expertise specialists regarding the use of blood products. In spite of this, no effective measures such as a switch to cryoprecipitate were taken, and the authorization of safe heated products was also delayed, as a result of which 40 % of Japanese hemophiliacs, or some 2,000 people, fell victim as a result of 'human error' [2]. Additionally, since insufficient risk data was provided, the infection spread to partners, families, and other associates of hemophiliacs through secondary and tertiary infection. In connection, questions were asked as to the degree of responsibility of those institutions involved in the outbreak and spread of the infection.

The cause of the outbreak has been said to lie in the web of collusive relationships linking industry, government, and academia and three groups in particular: the pharmaceutical manufacturers who undertook the manufacture and sale of the blood products, the former Ministry of Health and Welfare, which held approval and 'licensing authority' over these pharmaceutical products, and the AIDS Research Group commissioned by the former Ministry of Health and Welfare. Underpinning the collusion between these three groups must have been an 'unwritten law' recognized among them as inviolable. These point up the pressing need to reform the poor practice and customs inherent in Japan's pharmaceutical manufacturing industry and by extension the ethos and culture of Japan's industry organizations and management. This chapter aims to analyze the iatrogenic AIDS problem as a 'human-made disaster' [3] from a global perspective through international comparisons of the number of people infected with HIV, and also to look

---

Reworking of: Atsuji, S., *Management Policy for Organizational Disaster*, Doshisha University, 2003.

S. Atsuji, *Unsafety*, Translational Systems Sciences 7,
DOI 10.1007/978-4-431-55924-5_5

from a ‘socio-biological perspective’ at the possibility of viral outbreaks in the near future and the conditions that have precipitated this hazard.

5.1 AIDS Pandemic as Human-Made Disaster

5.1.1 Worldwide HIV Hazard

It is now over 30 years since the first AIDS case was reported in Los Angeles in the United States in 1981, after which AIDS began to spread worldwide. As of the end of 2011, the number of HIV-positive people worldwide had reached 34 million. In response to this situation, the United Nations released its *2011 Political Declaration on HIV and AIDS: Intensifying Our Efforts to Eliminate HIV and AIDS*. As of 2012, 186 countries or 96 % of United Nations members had submitted a report on their national AIDS-response measures (UNAIDS) [4]. Figure 5.1 indicates the size of HIV-positive populations, showing from left to right the infection rates among adults, sex workers, men who have sex with men, and people who inject drugs. It is striking that the HIV-positive rate among drug injectors is 22 times higher than among the general population.

The background to this situation regarding HIV infection is a story of massive spread of infection through an accident with ‘high-risk biotechnologies’ [5]. The iatrogenic HIV infection scandal refers to a series of cases of infection with the human immunodeficiency (HIV) virus, caused by the failure to apply safety

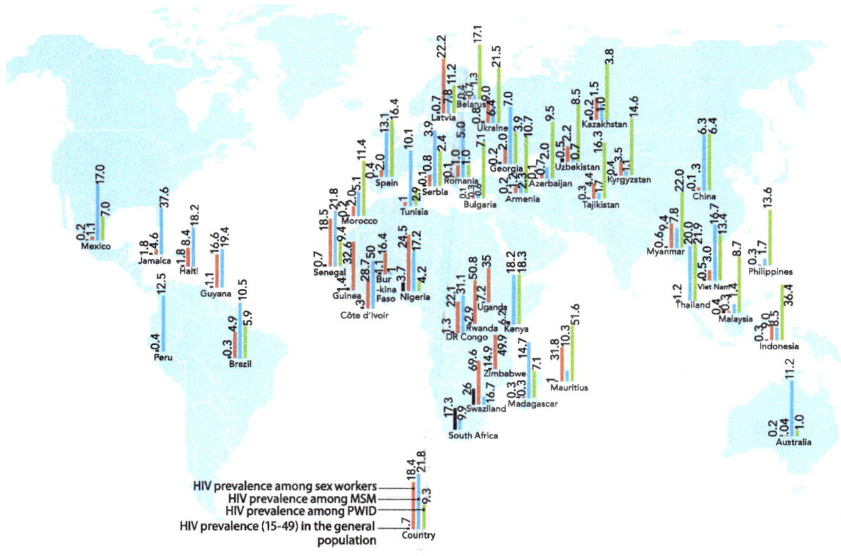

Fig. 5.1 HIV prevalence in adults and key populations 2012 *Source:* UNAIDS, *World AIDS Day Report*, United Nations, 2012, pp. 42–43 (*Publicity*)

measures to blood products. Included in these blood products were coagulation-factor products, which are pharmaceuticals effective against hemophilia. The raw material for these products is prepared from the blood of many donors, and the lack of a procedure for virus inactivation left no defense against viral infection. The reported discovery of acquired immune deficiency syndrome (AIDS) in 1981 led to concerns over the risk of infection through blood products, but despite the warnings from many specialists, the use of these products continued and resulted in some 2,000 cases of HIV infection. According to a 1997 White Paper from the Ministry of Health and Welfare, of the 4,028 people with HIV as of February 1997, 1,872 had been infected by blood products. Meanwhile, of the 1,484 people who had developed AIDS symptoms, 641 had been infected by blood products. The reality was that almost half of those with HIV infection were victims of the scandal. This affair raised issues regarding the administration of pharmaceutical approval and licensing by the former Ministry of Health and Welfare and the culpability of the involved organizations and individuals in industry (pharmaceutical manufacturers); in government (the former Ministry of Health and Welfare); and in academia (the AIDS Research Group led by Dr. Z).

5.1.2 Spread of HIV Infection through Pharmaceutical Drugs

Iatrogenic HIV infection differs from many other cases of iatrogenic drug damage in that it is a 'progressive condition'. What this means is that even when use of the drugs stops, the progression of the symptoms does not. This is because iatrogenic HIV infection arises not from the drug's side-effects, as with most other cases of iatrogenic drug damage, but from viral contamination of the blood that is the raw material for the drugs (Osaka HIV) [6]. In the absence of effective therapy, HIV causes a gradual weakening of the immune system, leading to the onset of AIDS. As of July 1998, there had been 493 deaths from iatrogenic HIV infection, while the number infected was 1,434, of whom 631 had developed AIDS. The number of deaths continues to increase. The fact that many children were affected is another distinctive feature of iatrogenic HIV infection, making all the more urgent the task of limiting the damage and developing therapies. Of those affected, 14 % were 10 years old or younger, while 30 % were 15 years or below, and those aged 20 or below accounted for more than half of all victims. This is why, in the first report by the Research Group for Preventive Therapies for People with HIV Infection, the group's chief researcher, Professor K. Yamada, called iatrogenic HIV infection "one of the greatest medical tragedies of this century". The group's report is subtitled *The AIDS Scandal* [7].

Table 5.1 International comparison of HIV-infected populations: 1999

Country	HIV-infected population (1990)	HIV-infected population (2011)	Number infected by blood products (estimate)
United States	1,200,000	2,000,000	6,000–10,000
Canada	46,000	89,000	2,000
Japan	6,900	10,000	2,000
France	120,000	200,000	4,000
Spain	97,000	160,000	2,200
Germany	39,000	82,000	1,900
Italy	190,000	200,000	1,300
United Kingdom	30,000	120,000	1,200

Source: UNAIDS, *World AIDS Day Report*, United Nations, 2011 (*Publicity*).

Table 5.1 summarizes the iatrogenic HIV problem in terms of the number of HIV-infected people and the (estimated) number infected by blood products in the major developed countries. The number of people infected iatrogenically through pharmaceuticals is particularly high in the United States and France, indicating that the damage was spread through a combination of natural and human-made disasters. In many countries including Japan, court actions were brought against the government and pharmaceutical manufacturers. In Spain, where 90 % of blood-coagulant products were imports from the United States, it is reported that 82 % of the 2,700 hemophiliacs became infected with HIV (Osaka HIV) [8].

In the case of the 'non-heat-treated' coagulation-factor products which were the medium of infection, although there were safety issues involved, the fact remained that they were effective for many hemophiliacs and user-friendly. The coagulation-factor products had a stronger pharmaceutical effect than the cryoprecipitate preparations which were candidate replacements, and, when symptoms became severe, there were apparently cases in which only coagulation-factor products could elicit recovery. Moreover, self-injection of coagulation-factor products by hemophiliacs was authorized in February 1983, offering a greater degree of user-friendliness than cryoprecipitate, which in principle had to be administered in a hospital. As HIV research was less advanced than nowadays, many doctors and patients opted for the immediate benefits of pharmaceutical efficacy and user-friendliness in the face of an unknown future risk, resulting in the unforeseen human-made medical disaster of iatrogenic HIV infection.

Central government (the former Ministry of Health and Welfare) continued to assume the role of the supervision and guidance authority for pharmaceutical approval and licensing in its relationship with pharmaceutical manufacturers. Pharmaceutical manufacturers submitted detailed reports on risk data and other information to the Ministry of Health and Welfare, which provided them with guidance on the range of countermeasures. The former Ministry of Health and Welfare had collected information from overseas specialists, but this information was not communicated to medical treatment institutions, patients, or the general public. At the time of the first confirmed case, the Ministry tried to prepare a 'soft landing' for AIDS in Japan. Similarly, so as to avoid panic among the public and unequal treatment of companies, it continued to advocate a softly-softly approach. In court, too, the government went on to adopt the role of director and defender of the accused companies. In addition to this aspect, there was a strong significance attached to the fact that the disease was infectious, which meant that the damage spread through secondary and tertiary infection. Surprisingly, as the Ministry initially treated HIV as a public health and sanitation issue, it did not provide victims with medical treatment, relief or compensation, thinking of the infection rather as a social evil which needed to be rooted out and eliminated.

5.1.3 Japan's Iatrogenic AIDS Epidemic

Detailed information on the timeline of the iatrogenic HIV infection scandal has appeared on the website of Life AIDS Project and in a book by Hitoshi Sakurai entitled *Umoreta eizu hōkoku* (The Buried AIDS Report) [9]. The timeline presented below was prepared with reference to these sources and is based on factual information already released into the public domain.

Year. month	
1981.6	The US Centers for Disease Control and Prevention (CDC) reports first AIDS patient (a male homosexual)
1982.7	CDC reports AIDS cases in hemophiliacs
1982.9	CDC adopts the term acquired immune deficiency syndrome (AIDS)
1982.12	The US Food and Drug Administration (FDA) proposes switch to cryoprecipitate and removal of at-risk groups from blood-donor eligibility The US National Hemophilia Foundation (NHF) points out the risk from blood products and the efficacy of cryoprecipitate
1983.1	NHF issues preventive warning and urges switch to cryoprecipitate
1983.2	Ministry of Health and Welfare recognizes home treatment of hemophilia (self-injection) for public health-insurance purposes
1983.3	FDA authorizes manufacture of heated products by Travenol USA FDA recommends halt to blood donation by high-risk groups
1983.4	CDC director points to efficacy of cryoprecipitate and the existence of numerous cases of latent infection. Isolation of the virus that causes AIDS (at the time called LAV) is announced in the journal *Science*. (Confirmed as causative virus in 1984) FDA orders development of heated products as a measure against AIDS
1983.6	Ministry of Health and Welfare establishes an AIDS fact-finding research group (AIDS Research Group, led by Dr. Z) World Federation of Hemophilia (WFH) Stockholm Congress approves continued use of non-heated products
1983.7	Hemophiliac dies at Teikyo University Hospital

(continued)

Year. month	
1983.9	National Hemophilia Network of Japan (a hemophilia-related association) requests the Ministry of Health and Welfare to speed the provision of heated products
1984.2	Travenol Japan begins clinical study of heated products
1984.6	Green Cross Corporation begins clinical study of heated products
1985.3	AIDS Survey and Investigation Committee recognizes a homosexual patient at Juntendo University Hospital as Japan's first AIDS case
1985.5–6	Hemophiliac treated with non-heated products at Teikyo University Hospital becomes infected with HIV and dies in 1991
1985.7	Ministry of Health and Welfare gives block approval to heated products
1986.1	Green Cross Corporation launches sale of heated products
1986.4	Liver-disease patient is treated with non-heated products sold by Green Cross Corporation at general hospital in Osaka Prefecture. Patient dies in 1995
1989.5–10	Initial HIV-related court actions brought in Osaka and Tokyo
1996.2	Health minister Naoto Kan accepts government's responsibility and apologizes
1996.3	Settlement reached in initial HIV-related court actions
1996.8–10	District Public Prosecutors Offices of Tokyo and Osaka order arrest of Dr. Z, Mr. Y (former head of the Ministry of Health and Welfare's Biologics Division), and three successive presidents of the Green Cross Corporation
1997.3	First public hearing at district courts of Tokyo and Osaka
2000.2	Three company presidents are sentenced to between 16 months' and 2 years' imprisonment at Osaka District Court
2001.3	Dr. Z found not guilty
2001.9	Mr. Y, former head of Ministry of Health and Welfare's Biologics and Antibiotics Division, sentenced to 1 year's imprisonment suspended for 2 years

5.1.4 Background to Scandal

Hemophilia, a disease in which constituents required for blood clotting and hemostasis are lacking or deficient for genetic reasons, requires medical treatment to supply these missing blood-coagulation factors. Cryoprecipitates products were used as drugs to control the disease [10]. As cryoprecipitate products are made from the pooled blood of one or at the most ten individuals, it was expected that they would be less likely to cause infection than coagulation-factor products, which are made from the pooled blood of between 2,000 and 20,000 people [11]. As the first cryoprecipitate were frozen products, their storage and use presented challenges, and patients had to be treated in hospital, but this problem was resolved by the dried cryoprecipitate developed in the early 1980s, and home treatment using cryoprecipitate did apparently take place in some cases. In countries such as Norway, which actually switched to cryoprecipitate, the rate of hemophiliac HIV infection was kept to low levels, indicating effective decision-making on health policy. This fact casts doubt on the basis of Dr. Z's statement that 'cryoprecipitate is liable to solidification', and instead points to the difference in drug-price markup as one of the possible reasons why the switch to cryoprecipitate was not realized.

From 1984 until around 1985, many Japanese doctors viewed as specialists and many officials at the Ministry of Health and Welfare were apparently dominated by the 'logical argument' that "although use of blood products can lead to HIV infection, many patients have not developed symptoms. Infection and onset of symptoms are thus separate phenomena. Although 'non-heated products' may be a cause of HIV infection, they do not act to cause AIDS onset" [12]. AIDS is a disease of immune dysfunction caused by a virus called HIV. In a document published in 1986, Dr. Z maintained that confirmation of AIDS requires three conditions to be fulfilled [13]. The first was infection with the AIDS virus, which could be tested via the presence of antibodies. The second was reduced or insufficient immunocompetence. The fact that this reduction of immunocompetence occurs over time is one of the characteristics of AIDS. For reduced immune function to result in the onset of symptoms can take more than 10 years in many cases, until which time the patient may experience no more than a certain degree of fatigue, with no impairment of everyday activities. This is still one of the features that distinguish HIV infection from the onset of AIDS. The third of Dr. Z's conditions was opportunistic infection, which appears as a result of weakened immunocompetence in the form of illnesses such as pneumocystis carinii pneumonia and Kaposi's sarcoma. As a result, medical treatment following the onset of AIDS consists of a constant battle with opportunistic infections.

Figure 5.2 shows the number of new HIV and AIDS infections and deaths in the two subsequent decades from 1990 to 2010. Many HIV-infected patients die each year; the total number of those infected is shown in the figure. When we consider also the number of AIDS deaths due to non-pharmaceutical-related natural infection, such as infection through unprotected sex or needle-sharing, the extent of the tragedy is immeasurable. The problem of iatrogenic AIDS thus continues into the present.

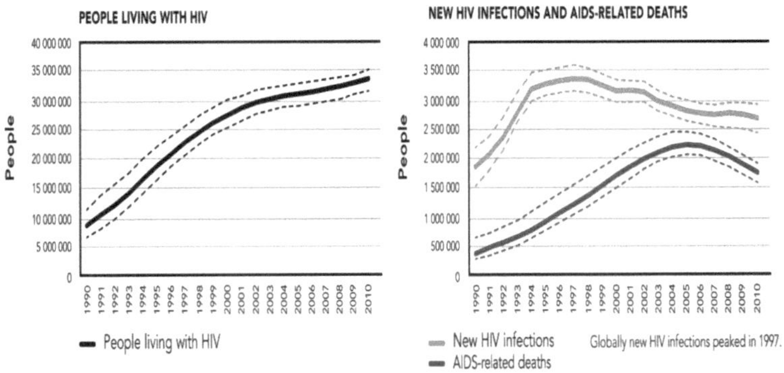

Fig. 5.2 People living with HIV around the world *Source:* UNAIDS, *World AIDS Day Report*, United Nations, 2011, pp. 6–7 (*Publicity*)

In the case of iatrogenic HIV infection, the main point of legal contention was at what juncture pharmaceutical manufacturers and the Ministry of Health and Welfare had become aware of the risk from non-heated products. The case in point in the criminal trial was that of a liver-disease patient treated with non-heated products from the pharmaceutical company Green Cross Corporation in April 1986. The court attempted to trace why the president of the corporation had not recalled the products, and why Mr. Y, the then head of the Biologics and Antibiotics Division at the former Ministry of Health and Welfare, had not ordered a recall.

5.2 Systemic Breakdown due to Ghost Governance and Lost Compliance

5.2.1 Business Ethics and Corporate Social Responsibility

Since pharmaceutical manufacturers were also in a position to access risk data, they were unable to deny their involvement in the manufacture and sale of drugs potentially contaminated with the virus. Information from pharmaceutical manufacturers was reported to the former Ministry of Health and Welfare, but was not communicated to the hemophiliacs who were the product consumers. The Ministry thereby clearly neglected its duty to warn patients and the public. As for non-heated products, while most other countries reduced their imports, Japan sought to increase imports and even reduced the price as an incentive. The reason for the import of unsafe 'non-heated products' was quite simply the prioritization of earnings from the markup on the drug price. This is the profit derived from the difference between the fixed domestic price for medical products and the price actually paid to the overseas supplier. The markup on non-heated products was particularly great. With cryoprecipitate products, in contrast, there was almost none. There are also issues associated with the actions of those involved after the danger was realized: even after heated products became available, they failed to urgently recall the non-heated products which they knew to be unsafe. By making false reports to the Ministry of Health and Welfare regarding the dates of market release and recall of non-heated products, the Green Cross Corporation engaged in fraud and misrepresentation compounding the damage, for which it bears a grave 'social responsibility.'

Figure 5.3, stakeholders of the pharmaceutical industry, presents a diagrammatic illustration of the interest groups in the industry involved in the administration of pharmaceutical approval and licensing. As shown, Green Cross effectively engaged in bribery, which included making payments, disguised as 'research funds', to a foundation that Dr. Z was in the process of establishing, suggesting that the 'business ethics' of this pharmaceutical operator were very poor. Its approach, which prioritized its own interest at the expense of patients and the public, aggravated the damage. However, this kind of poor practice was not restricted to the

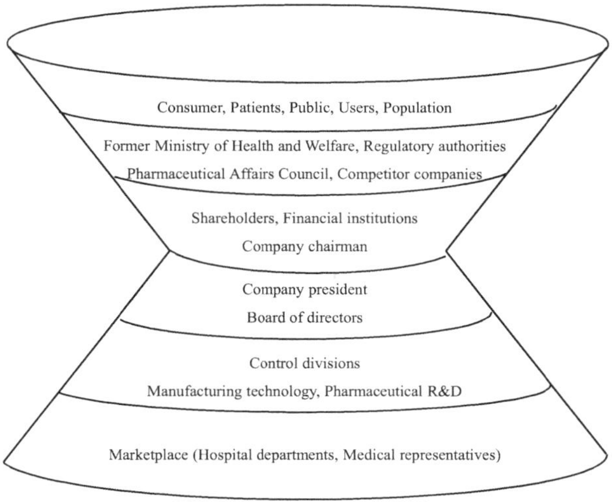

Fig. 5.3 Stakeholders of the pharmaceutical industry. *Note:* Drawing of Japanese pharmaceutical industry based on the concept of 'Stakeholders and Governance' by A.A. Berle

corporate players; responsibility also lies with the government authorities that administered the pharmaceutical regulatory system, which became ever clearer as the Japanese media (NHK) delved more deeply into the scandal.

The former Ministry of Health and Welfare had the role of providing supervision and guidance to pharmaceutical manufacturers. Notwithstanding its name change to the Ministry of Health, Labor and Welfare following a ministerial reorganization, its historic responsibility for the iatrogenic HIV infection remains, together with its duty to compensate patients and victims. This raises the question of the political and administrative responsibility of the regulatory authorities. Although it may not have been deliberately responsible, if it is the case that the former Ministry of Health and Welfare sacrificed the health of hemophiliacs for the sake of pharmaceutical manufacturers' profit, then the regulatory authority itself has become a malignant influence on society, a situation that calls for the establishment of an ethical code for government organizations. The pharmaceutical regulatory system has been called a 'web of vested interests'. The sorting of good medicines from bad is carried out under government guidance through a process known as pharmaceutical inspection, but, paradoxically, there is greater interest vested in bad medicines than in good medicines. Pharmaceutical companies make massive capital investment in drug development and have to adapt to new raw materials and a new production line for each new drug. They need to recoup this investment, and the development of a single drug is thus a kind of venture project that can determine the entire fate of the company.

Because of this, once a drug had been developed, since it was no longer possible to halt the investment in the associated production plant, the company would approach official bodies and the specialist committees and research groups that had the effective decision-making authority within the pharmaceutical-inspection system of the former Ministry of Health and Welfare with the intention of having the drug approved, regardless of its merit. It was this network of interests surrounding pharmaceutical regulation that formed the backdrop to the tragedy of iatrogenic HIV infection. Meanwhile, the inspection procedures and clinical studies involved in drug licensing under the government regulatory system require time, something which often stands in the way of delivering medical treatment to patients who need the drug. Indeed, one wonders what the purpose of the government regulatory system is, and whether the clinical studies involved in pharmaceutical inspection have any useful effect. The iatrogenic HIV infection scandal provoked a loss of confidence in the pharmaceutical regulatory system, and the former Ministry of Health and Welfare and initiated a fundamental questioning of the whole of Japan's medical-treatment policy.

5.2.2 *Ghost Governance*

The administrative responsibility of the regulatory authorities involved in the iatrogenic HIV infection scandal (the former Ministry of Health and Welfare) was the focus of the court case. In court, the basis for the questioning of the Ministry's responsibility was the Pharmaceutical Affairs Law, which was amended in 1979 in response to the SMON (subacute myelo-optico-neuropathy) case. The amendment gave the Ministry the authority to close down the supply of a pharmaceutical where there was the risk of it creating or aggravating a health or safety hazard. At issue was the failure to exercise this authority, in other words negligence on the part of the regulatory authority. Specifically, it can be pointed out, for instance, that Japan lagged behind the United States by 2 years and 4 months in the approval of heated products. This was largely attributable to the coordination of clinical studies by Dr. Z, but another probable issue was the structure whereby the former Ministry of Health and Welfare's pharmaceutical-licensing operation was dependent on a small number of specialists. Table 5.2 summarizes the dates from which HIV-antibody testing and heat treatments of concentrated products were made obligatory.

Documents now made public suggest that, although the former Ministry of Health and Welfare had collected a large amount of risk data, it not only did not make the data public but may actually have manipulated the advice from the FDA. AIDS was first recognized in Japan on March 22, 1985, with the reporting of the case of a male homosexual resident in the United States, which became known as the Juntendo University case. The patient survived for 10 years after diagnosis, and, although infected with HIV, is thought to have been yet to develop AIDS at that time. In contrast, a hemophiliac who died in July 1983, known as the Teikyo University case, was not recognized as an AIDS patient. It has been claimed that

Table 5.2 Date of introduction of compulsory HIV-antibody testing and heat treatment

	Introduction of obligatory HIV-antibody testing (ELISA)	Introduction of obligatory heat treatment of concentrated product
Australia	May 1985	January 1985
Canada	November 1985	July 1985
Denmark	January 1986	October 1985
France	August 1989	October 1985
Germany	October 1985	Instead of obligatory heat treatment, no reimbursement for 'non-heat-treated' products
Italy	March 1985	July 1985
Japan	November 1986	October 1985 to February 1986
Britain	October 1985	June 1985
United States	March 1985	October 1984
India	1987–1989	N/A
Thailand	1987–1989	N/A
Zimbabwe	July 1985	N/A

Source: E. Feldman and R. Bayer, *Blood Feuds*: *AIDS, Blood, and the Politics of Medical Disaster*, 1999, p. 341

this was because the former Ministry of Health and Welfare tried to conceal the fact of hemophiliac HIV infection. Even subsequent to the licensing of heated products, the Ministry exacerbated the damage by failing to order pharmaceutical manufacturers to recall unheated products. Underlying this lax leadership is a readily perceptible collusion between the former Ministry of Health and Welfare and the pharmaceutical manufacturers. The Green Cross Corporation of those days was described as an outpost of the Ministry of Health and Welfare's Pharmaceutical Affairs Bureau, and the practice of *amakudari*, whereby officials from the Ministry frequently retired into sinecure positions with pharmaceutical manufacturers, led to the formation of strong collusive relationships in the area of pharmaceutical approval and licensing (Mainichi Shimbun Shakaibu) [14].

As shown in Fig. 5.4, the background to the iatrogenic HIV infection scandal was formed not only by the various contributions of industry, government and academia, but also by the undeniable fact of mutual interorganizational relationships among the three based on an 'industrial protection policy' peculiar to what has been called *Japan Inc*. It was surely the collusive relationships among industry, government, and academia based on the 'vulnerable systems' in Japanese society [15]—a negative result of the interrelationship of organizations in Japan's industrial society—that lay behind the iatrogenic HIV infection scandal.

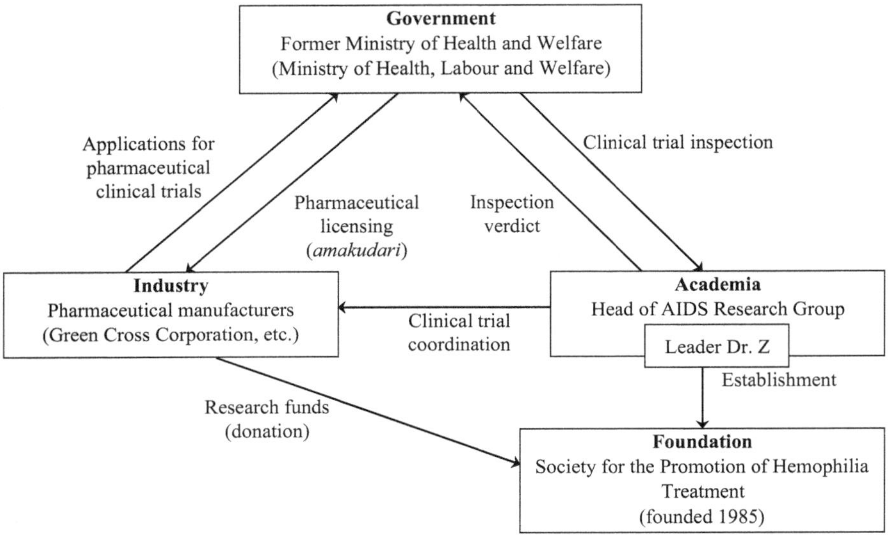

Fig. 5.4 Lost compliances: illegal collusive relationships. *Note:* Charting based on 'Organizational Inertia and Culture' by C. Argyris, J.G. March, G.H. Litwin, and R.A. Stringer

5.2.3 Lost Compliance

The head of the AIDS Research Group, Dr. Z, was at the time Japan's foremost expert in hemophilia treatment, and was thus an influential and authoritative voice in the field of medicine relevant to the iatrogenic HIV infection scandal. The AIDS Research Group was in fact heavily involved in important decisions that led to the compounding of the damage from the iatrogenic HIV infection, such as the decision not to switch from coagulation-factor products to cryoprecipitate products. Specifically, at the meeting of the Home Treatment Promotion Committee on October 18, 1983, Dr. Z said in respect of his earlier claim that 'cryoprecipitate tends to solidify' that it might not be true. At a time when there were doubts over the safety of blood products, a grave responsibility attaches to his decisions not to switch to cryoprecipitate products, and then to delay the clinical study of the heated products when he should have been working to provide a safer drug as quickly as possible. In connection, it was found that doctors had had financial dealings with pharmaceutical manufacturers, which was proven by the fact that Green Cross had made donations, disguised as research funds, to a foundation established by Dr. Z.

As part of the July 1997 reorganization of the pharmaceutical regulatory structure recommended in the 1997 Annual Report on Health and Welfare, the Pharmaceutical Affairs Bureau was abolished, and the Health Policy Bureau took control of policy on the promotion of pharmaceutical research and development, pharmaceutical production and distribution, and related aspects. Meanwhile, the Pharmaceutical and Food Safety Bureau was created to oversee the broad range of safety measures relating to medical treatment and pharmaceuticals, such as pharmaceutical clinical studies, approval and inspection, post-marketing safety measures, and measures to prevent infection within medical-treatment institutions. In parallel, as part of a strategic upgrade and reorganization of testing and research institutions, the National Institute of Hygienic Sciences was reconstituted as the National Institute of Health Sciences, and was equipped with a Pharmaceuticals and Medical Devices Evaluation Center to promote survey and research work on the safety and efficacy of pharmaceuticals and foodstuffs and strengthen pharmaceutical inspection. To coincide with the reorganization, there was a systematic increase in the number of staff assigned to pharmaceutical inspection work, and the inspection system was changed: whereas previously the actual work of inspection had been delegated to the Central Pharmaceutical Affairs Council, whose members were drawn from the outside, the core work of pharmaceutical inspection was now to be carried out in-house by the Pharmaceuticals and Medical Devices Evaluation Center, with the Council's role specialized to cover high-level assessment. Urgent steps were to be taken to set up a corresponding system of responsibilities. The system of clinical-study principal investigators, which had been Dr. Z's role in the licensing of heated products, was abolished.

The new medical-product approval and inspection system created by the reorganization of government agencies came into force in 1997. In subsequent administrative reforms, the former Ministry of Health and Welfare became the Ministry of Labor, Health and Welfare, and there were changes in the content of its administrative activities including services provided and approval and licensing operations. Under the previous system, Inspection Guidelines and Inspection Reports were submitted to the Investigation Committee of the Central Pharmaceutical Affairs Council, which was made up of specialists from outside the Ministry. This system was replaced in November 2000 by a simplified format in which the Central Pharmaceutical Affairs Council participated in a so-called team inspection. This move appears to have been designed to shorten the time required for approval.

In the final analysis, the cause of the iatrogenic HIV infection lay in a system in which the government agencies that constituted the pharmaceutical regulatory authorities were essentially bypassed so that the effective 'decision-making' body for drug clinical-study approval was the specialist committee known as the AIDS Research Group, to which pharmaceutical manufacturers made donations in an effort to have their pharmaceuticals approved and licensed, thus interfering in the process of pharmaceutical approval and license renewal. In future, lessons must be drawn from the iatrogenic HIV infection, and reform must be applied to the web of interests in the pharmaceutical industry and its system of cozy collusive relationships with government agencies. The poor practice and collusive relationships arising from the network of interests linking the worlds of industry and government also exemplify the need to reform, in terms of both policies and systems, the ethos and culture which became entrenched during Japan's earlier period of rapid economic growth. An existing organizational culture must not be allowed to prevent the organization from learning to find its way in a new age. Dismantling the old culture of the past and creating a new culture is an important process. The culture of poor practice which is latent within the pharmaceutical regulatory system and the pharmaceutical industry represents a form of 'organizational inertia' which still today retains the power to act against the public interest. The culture and ethos latent in this network of collusive relationships among Japanese industry, government, and academia may need to be the subject of creative destruction as we move into the near future.

It is clear that responsibility for the organizational disaster of iatrogenic HIV infection lies at once with industry, government, and academia, but the background to this is the industrial-protection policy of *Japan Inc.*, which is deeply rooted in the culture of the pharmaceutical industry. The system for approval and licensing, as administered by the regulatory government authority for each industry, itself spawned a network of collusive relationships among industry, government, and academia through practices such as *amakudari*, donations from private-sector bodies to the public sector, and expense-account entertaining. The case of iatrogenic HIV infection can be seen as a negative result of this relationship between business and government. This underlines the fragility of the 'cultural ethos' [16] of Japanese industrial organizations and typifies a system common to all of Japanese industrial society: a veritable 'Galapagos, a strange, isolated world'.

5.3 Viral Outbreaks Caused by Global Warming: Limitations of Management and Policy

Human-made disasters such as the phenomenon of iatrogenic AIDS infection outlined above have an inevitable 'teleconnection' with natural disasters. New viral strains and infectious diseases are one form of 'unsafety threatening humankind.' Historically, tuberculosis, cholera, plague, influenza, AIDS, and other infectious diseases have claimed many victims. More recently, the worldwide spread of SARS and avian influenza in 2003 is fresh in our memory. There are fears of epidemics of the three major infectious diseases (AIDS, malaria, and tuberculosis) arising from the 'biological hazard' caused by the world population explosion to 7.2 billion. The particular danger of biological hazard is its exponential pattern of spread. Explosive damage arises through bacterial and viral infection via living organisms, person-to-person infection, or cross-species infection, for example from cattle to humans through bovine spongiform encephalopathy (BSE), variant Creutzfeldt–Jakob disease (vCJD), and foot-and-mouth disease. Moreover, rubella infection in pregnant women may cause cataracts and glaucoma, congenital heart disease, hearing impairment, and other conditions in the fetus, while congenital rubella syndrome (CRS) poses the risk of damage to the next generation. Whether this is regarded as a case of human-made or natural selection, the result is the same: the disaster leads to an indivisible resonance phenomenon, the negative interaction of which causes the accelerated spread of 'unsafety'.

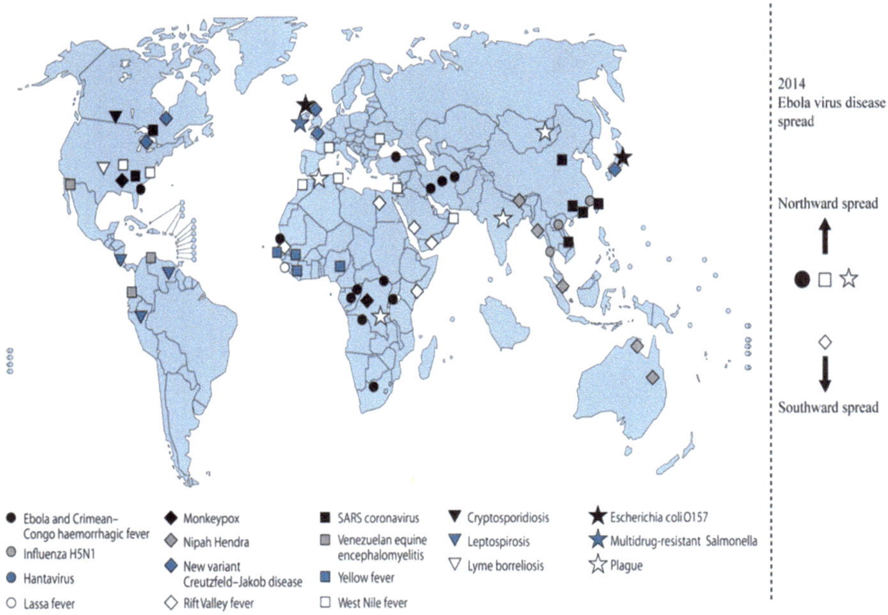

Fig. 5.5 Selected emerging and re-emerging infectious diseases: 1996–2004 *Source:* WHO, *World Health Report 2007: A Safer Future: Global Public Health Security in the 21th century*, p. 12 (*Publicity*)

The WHO has published Fig. 5.5 worldwide to signal the risk from biological hazards. The northward spread of infectious tropical diseases caused by recent global warming is proceeding at an ever-accelerating pace. Among the hazards facing the world, the proportion represented by these biological hazards is second in number only to natural disasters such as earthquakes, tsunamis, volcanoes, typhoons and hurricanes, and accounts for one-third of all hazards.

Meanwhile, there is concern that global warming's disruption of the energy balance may allow the spread of infectious tropical diseases to the northern hemisphere. The serious prevalence of West Nile fever in the United States resulted from its being spread to the temperate zone of North America by travelers. In Japan, similarly, the example of the redback spider (*Latrodectus hasselti*), which was discovered in Osaka Prefecture and has extended its habitat to the whole country, demonstrates that this is not an insubstantial problem. In particular the recent prevalence of Dengue-fever, whose incidence has increased 30-fold in the last 50 years, means that over 100 countries are threatened by the growth of the domain of infection (WHO) [17]. Figure 5.6 shows areas with high risk of Dengue-fever infection as of 2011. The lines to the north and south of the figure indicate the minimum temperature of 10 °C delineating the habitat limit of the mosquito that transmits the Dengue-fever virus. Advancing northward like an army, global warming is extending the habitat of the mosquitoes that transmit tropical viruses

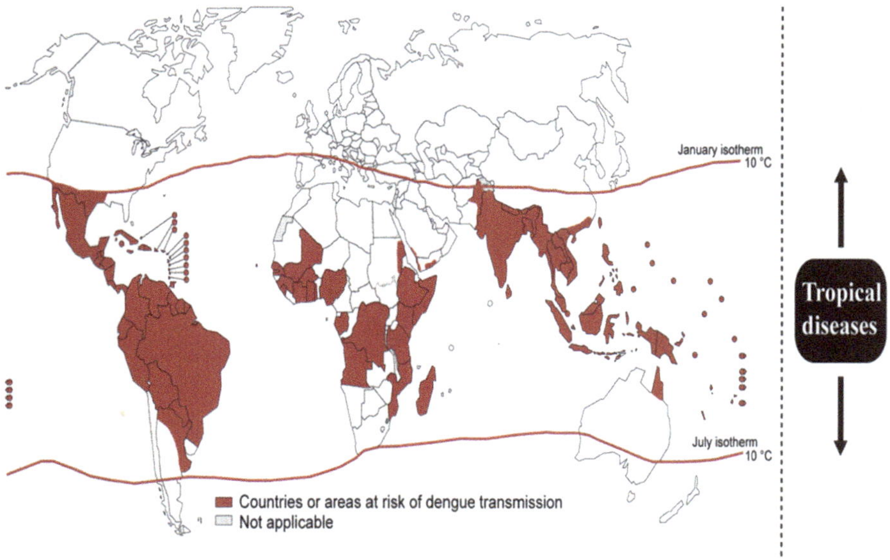

Fig. 5.6 Global distribution of countries or areas at risk of Dengue transmission, 2011. *Note:* Mapping based on the discussions with participants at *Climate Change and Global Warming* by WWF Japan 2011 (Adapted from "Sustaining the drive to overcome the global impact of neglected tropical diseases", *WHO*, 2012, p. 25)

and is pushing the infection toward the northern hemisphere, which has a large land mass and is home to a large proportion of the human race. This poses a threat to the populations of these areas, who have no experience of or resistance to tropical viruses, diehard by global-warming.

Not only the abovementioned infectious tropical diseases transmitted by bacteria and viruses, but also the risk of infection spread through global warming and the resulting crisis, will be of increasing concern going forward. For instance, since cholera bacteria live in symbiosis with plankton in seawater, the rise in sea temperatures, causing plankton to breed more prolifically, also leads to an increase in cholera bacteria, which has extended its infection zone northward. Already, it is reported that the 1991 El Niño phenomenon in South America has led to sharp year-on-year increases in cholera cases. The IPCC report from the end of September 2013 states that the world's average atmospheric temperature rose by 0.85°C from 1880 to 2012 and predicts that the temperature rise by the year 2100 will be up to a maximum of 4.8 °C, leading to fears of a 'Global Big Melt', which will precipitate a worldwide struggle over water and food resources (IPCC) [18]. The freshwater available on the planet for human consumption as drinking water is said to represent 0.008 % of all the earth's H_2O, so a Global Big Melt would mean the depletion of the water resources for the human population of 7.2 billion now. Especially in the northern hemisphere, which contains a high proportion of the planet's land mass, the melting of glaciers and permafrost soil to which the Big Melt refers is predicted to lead to the spread of viral infection to previously unaffected areas. Combined with 'trans-global movement' of travelers and migrants, and biological weapons, terrorism, and other disasters arising from human-made unsafety, these outbreaks could spread worldwide.

Governments and the responsible departments of regulatory authorities are loath to recognize 'socio-biological hazard'. The leak and spread of radioactivity following the meltdown of the Fukushima nuclear power plant, although a question of life and death for local residents, the wider community, and the Japanese population as a whole, were hidden by the government and the company involved. Socio-biological hazard thus cannot be controlled by central government policy or corporate management, which instead frequently responds with concealment or falsification of information. Nor is it susceptible to control by social or other systems. Outbreaks or pandemics of social or biological problems are accompanied by the breakdown of social functions, indicating the limitations of policy and management at the level of national government and business organization.

Unlike war, coups d'état, and conflict, the influx of people into an area of hazard results in new infections as contamination with the pathogen spreads along the chain among the members of families, communities, and organizations. National governments and the WHO have, albeit discreetly, sounded the alarm over the worldwide spread of locally endemic diseases not only through mosquitoes, ticks, and migratory birds, but also through human movement (travelers on business or otherwise). A crucial role in the infection zones has been played by the organization Médecins Sans Frontières, known for its role in the discovery of SARS. In such a spread of infection, as in the model predicted by J. Reason, [19] accidents and disasters leak through security holes, author which suggests a resonance between human-made and natural disasters.

To summarize, the increased risks and crises brought about by global warming can emanate through leaks in physical, social, and biological defensive barriers. The resulting human-made disasters have already brought about systemic breakdown on various fronts. The 'survivability' which is a defensive barrier programmed into human DNA does not operate in C. I. Barnard's so-called 'zone of indifference', where hazard is neither made known nor perceived. Consequently one could suggest, in many cases, hazard is only registered when a crisis emerges from the damage due to the spread of infection instigated in the breakdown of health and sanitation and other social systems and functions (Fig. 5.7).

A comparison of viral outbreaks on a global scale, such as the worldwide pandemic of iatrogenic AIDS, reveals a similar structure. First of all, insufficient information disclosure allows an influx of people into the infection area; secondly, government measures to suppress infection and efforts at an organizational level by corporations or other bodies remain weak; and thirdly, there is a "zone of indifference outside the infection area." These three factors create disregard for hazard information (de-civilization). Accurate publicity of 'socio-biological hazard' is therefore essential at the levels of international society, government, corporate organizations, and the individual, while also urgent are preventing unnecessary or unauthorized business visits or travel to the hazard area and other issues of

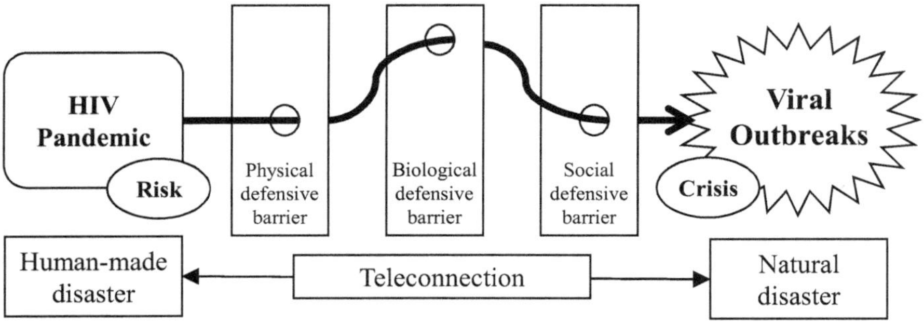

Fig. 5.7 Mechanism of socio-biological hazard. *Note:* Charting based on 'Security Holes' in Organizational Accidents by J. Reason and 'Systems Pathology' by L. Troncale

organizational compliance and governance, together with Human Resource Management (HRM). The northward spread of tropical infectious diseases through global warming has created a need for social systems at various levels, including those of government, corporate organizations, and the individual. This means that the concept of an 'eco-civilization'—promoting coexistence at the level of the social ecosphere, and associated disclosure of information—is the only viable approach to suppressing the combination of human-made and natural disasters that constitute socio-biological outbreaks and pandemics.

5.4 Postscript for Executives and Administrators

In April 2002, eastern Asia was struck by the SARS virus. Following the noble efforts made by the organization Médecins Sans Frontières, alerts were communicated worldwide through WHO. At the time, I was due to leave for a period of external research at IMD Lausanne, and my office urged me to proceed with the departure from Kansai Airport despite the risk of spreading the SARS infection. However, I defied them and postponed the departure. It was precisely on the day of my scheduled departure that Kansai Airport was subjected to a major disinfection operation as a precaution against SARS. The decision to postpone my departure after consulting my departmental head was a close call. If I had left on that day, I may have spread the infection to my research host institution at the IMD, and the SARS infection could have been transmitted after my return to many of my students and teaching colleagues. The conclusion to be drawn from this episode is that, going forward, whatever the field of work and in the management of organizations of all kinds, response to viral outbreaks will become an urgent task for participants in and administrators of business travel. I pray for the repose of the souls of those who died of SARS and influenza.

The recent hostage killings in Algeria and the deaths of journalists in Syria serve to illustrate that the managerial staff who issue the order for overseas business trips not only have managerial responsibility for the individual organization, but may also bear lifelong moral responsibility and a duty to compensate the families of injured junior staff for their trauma and emotional suffering. Accordingly, it has become essential for modern management to prepare for socio-biological hazard in organizational management by providing managerial staff with training in *risk, crisis and resilience management*. Reviewing the ups and downs of the Japanese economy, one recalls that, after the collapse of the bubble economy, a large portion of the sharply increased number of suicides from overwork that resulted from mass layoffs and staff cuts was represented by managerial staff which had fired their colleagues.

Coincidental though it may be, I was in the countries at the time of the military coup d'état under the Fujimori government in Peru and the coup d'état at Bangkok airport in Thailand. The Great East Japan Earthquake, the Hanshin-Awaji earthquake, and the death of a student in a Japan Railways accident are also among my various experiences of accident and disaster. It was because of these that I entered my present field of research with the aim of averting suffering caused by avoidable human-made disasters. This chapter is dedicated to the world, to its people, and to humankind as a whole.

References

1. Atsuji, S., *Management Policy for Organizational Disaster*, Doshisha University, 2003.
2. Reason, J., *Human Error*, New York: Cambridge University Press, 1990.
3. Turner, B., *Man-made Disaster*, London: Wykeham, 1978.
4. UNAIDS (2012), Global report: *UNAIDS Report on the Global AIDS Epidemic 2012*, 2012, pp. 6–8.
5. Perrow, C., *Normal Accidents: Living with High-Risk Technologies*, New York: Basic Books, 1984.
6. Osaka HIV Sosho Bengōdan, *Yakugai AIDS Kokusai Kaigi [Medically Induced AIDS International Conference]*, Sairyūsha, 1998, p. 161.
7. *See, The AIDS Scandal,* Iwanami Booklet. *See also*, URL: http://www.t3.rim.or.jp/~aids/yakugai2.html
8. Eric, F. and Ronald, B., *Blood Feuds: AIDs, Blood, and the Politics of Medical Disaster*, Oxford University Press, 1999.
9. *See, Life AIDS Project* (http://www.lap.jp/lap2/data/yakugain.html).
10. Kanuma, K., *Yakugai AIDS Saikō [Rethinking AIDS]*, Kadensha, 1998, p. 21.
11. Kanuma, K., *Ibid*., p. 22.
12. Kanuma, K., *Ibid*., p. 148.
13. Abe, H., *AIDS to wa Nanika [What is AIDS?]*, NHK Publishing, 1986, p. 28.
14. *See for example,* Mainichi Shimbun Shakaibu, *Yakugai AIDS Ubawareta Mirai [Medically Induced AIDS: Stolen Future]*, Mainichi Shimbunsha, 1996, p. 74.
15. Mintzberg, H., *Mintzberg on Management: Inside Our Strange World of Organizations*, New York: The Free Press, 1989.
16. Weick, K.E., "The vulnerable system: an analysis of the Tenerife air disaster" in Frost P.J. *et al.* (eds), *Reframing Organizational Culture*, London: Sage Publications, 1991.

17. WHO: World Health Organization, *Department of Control of Neglected Tropical Diseases*, 2012, p. 26.
18. IPCC: Intergovernmental Panel on Climate Change, *Climate Change 2013: The Physical Science Basis*, 2013. *See*, Weizsacker, E.U. von, Hargroves, K., and Smith, M., FAKTOR FUNF: Die Formel fur nachhaltiges Wachstum, The Natural Edge Project, 2009. *See also*, "Climate Change Act 2008" by UK Law.
19. Reason, J., *Managing the Risks of Organizational Accidents*, Ashgate Publishing, 1977, pp. 11–13.

Chapter 6
Boiling Globe: Cumulative Thermal Effluent from the World's 441 Nuclear Reactors over 40 Years

Hiroyuki Itsuki has said that Fukushima was a 'second war defeat'. Japan, which suffered the atomic bombing of "Hiroshima and Nagasaki" in World War II, was once again visited by a nuclear incident at Fukushima. After the World War, the state was defeated but the natural environment was preserved. Conversely, at Fukushima, the natural environment was lost and people were robbed of their livelihood, with the state alone remaining intact [1]. Historically, the International Atomic Energy Agency (IAEA) have taken only retrospective action in the event of nuclear-related accidents, disasters, or mishaps, while current law is insufficient and ineffectual in the face of the nuclear issue. Meanwhile, the management of the electric-power companies in charge of nuclear operations, such as the Tokyo Electric Power Company (TEPCO) in the case of the Fukushima nuclear accident, has also been lax both in its preventive measures against accidents and disasters and in its risk awareness [2]. Even after the accident, its response can only be called inadequate.

The present chapter (1) outlines the 'unstoppable nature' of nuclear generation as exemplified by the life cycle of nuclear reactor technology, the decommissioning of reactors, the nuclear radioactive wastes, and 441 reactors disposal coolants problems; (2) traces the roles in the JCO nuclear-fuel criticality accident of failed management in the form of the power companies, and government in the form of the "nuclear-electricity regulatory authorities and fuzzy policy"; and (3) highlights 'ocean-temperature' rise in the northern hemisphere, specifically the North Pacific, Arctic and North Atlantic, as a result of atmospheric global warming related to hydrosphere warming from the accumulative effect over 40 years of thermal effluent 'coolant water' from the world's 441 nuclear reactors, a negative heritage of the nuclear industry.

Reworking of: Atsuji, S., "Un–safety: Systems Pathology of the Fukushima Nuclear Catastrophe", *ISSS Proceedings*, 2013 and Atsuji, S. et al., "Sustainable Decision–making Following the Fukushima Nuclear Catastrophe", *IFSAM*, 2012.

S. Atsuji, *Unsafety*, Translational Systems Sciences 7,
DOI 10.1007/978-4-431-55924-5_6

6.1 Limits of Crisis Management concerning Aging Reactors

6.1.1 Systemic Life Cycle of a Nuclear Power Station

The disaster that occurred in March 2011 at the Fukushima nuclear power station in Japan sent shock waves around the world. With this now the third major nuclear power disaster, following Three Mile Island in America and Chernobyl in the former Soviet Union [3], the safety of nuclear power has begun to be questioned. In Western countries and other developed nations that have introduced nuclear power, the disaster has raised the issue of 'aging nuclear reactors', whose environmental impact, including the issue of decommissioning of nuclear reactors, has become a concern.

Of the world's 441 nuclear power reactors, 435 are concentrated in the northern hemisphere, leaving only six in the southern hemisphere. According to Dr. Koide of Kyoto University, the volume of the resulting thermal effluent water in the form of nuclear reactor coolants from a plant in the average "1-million kW output range is 70 tons/sec of thermal water", which has been heated by 7 °C. Assuming that the world's nuclear reactors operate at 70 % of capacity, with an average operating period to date of 31 years, it is estimated that "a cumulative total of 17.9 trillion tons of water has been heated by 7 °C." That is enough to form in the northern hemisphere a surface layer of around "11 cm that has been warmed by 7 °C, or a surface layer of 77 cm warmed by 1 °C."

Compared to atmospheric warming by CO_2, heat energy retained in seawater, because of the latter's specific heat, is more easily stored and less easily released, which poses the possibility that warming from nuclear power is causing a *global boiling* phenomenon in which the world's oceans, especially in the northern hemisphere, are overheating. According to the analysis of the research by author's KAKEN group funded by a Japanese government foundation, the possibility that the year-by-year accumulation of such thermal effluents from reactor coolants in nuclear power stations produces large bodies of water in the seas of the northern hemisphere while having a considerable influence on abnormal weather patterns arising from a process of teleconnection triggered by North Atlantic hotspots, cannot be excluded either qualitatively or quantitatively. This could be considered as like the 'environmental hormones' referred to by T. Colborn. It cannot therefore be ruled out that the 'human-made disaster' of nuclear-based thermal effluents, building up year by year, has precipitated 'global warming', abnormal weather patterns, and natural disasters such as summer blizzards, major floods, tornados, 'super-typhoons', and 'El Niño–La Niña phenomena' with the ironic result that "a chain of human-made disasters adds up to a natural disaster."

Figure 6.1 charts the world's nuclear power stations by duration of operation and shows a large number that have been operating for 30 years or more in North America, Europe, and Japan. Of those stations currently in operation, approximately 37 % are in the aging category—that is, 30 years old or more—in which

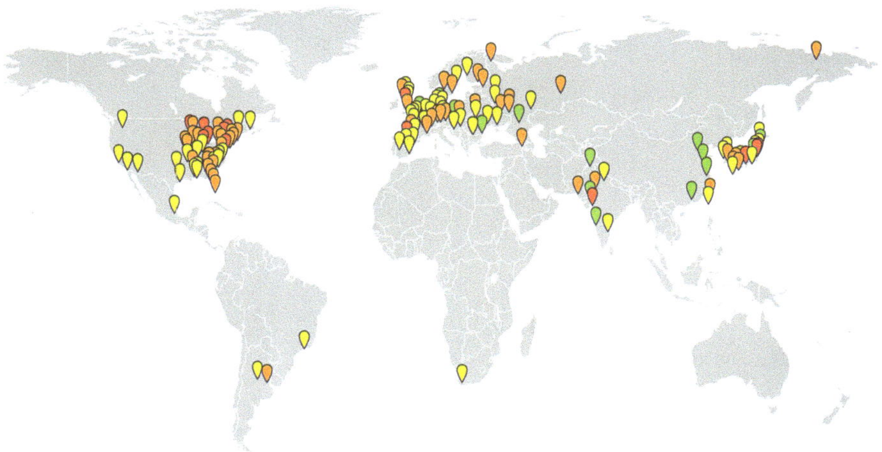

Fig. 6.1 The world's aging nuclear reactors (2011). *Note:* Mapping by R. Fujimoto and S. Atsuji based on Nuclear Database, World Nuclear Association

the life cycle has been extended beyond the normal operational life span of nuclear reactors. Meanwhile, standards for the decommissioning of nuclear reactors do not exist either at the international or the national level, and in the profit-driven and highly lucrative business of nuclear power generation, there is a history of operational life span being extended without allowing for 'decommissioning' and decontamination costs or accident clear-up costs. Calculations of costs have failed to consider expenditures and time periods falling outside the operational life span, at the planning and construction stage or in the dismantling and decommissioning of reactors. Table 6.1 summarizes the systemic life cycle of the nuclear power station including these stages. Normally, the life cycle of a nuclear power station has been set at 20 years, but many countries extend the operational life span beyond 40 years. The United States has permitted a 20-year extension of a nuclear power station already in operation for 40 years to a total of 60 years. Meanwhile, in Japan, the approval of extensions up to 60 years had been suggested in October 2010, the year before the Fukushima nuclear accident.

The period required for the decommissioning of nuclear reactors is said to be 40 years, which means that the life cycle from construction through to decommissioning, even excluding the disposal of spent nuclear fuel, is more than 80 years. The cost of decommissioning is estimated at around 350–480 million dollars for a small reactor (in the 500,000 kW range), around 430–610 million dollars for a medium reactor (in the 800,000 kW range), and around 560–760 million dollars for a large reactor (in the 1.1-million-kW range) [4]. Moreover, the planning and application process—from the establishment of a nuclear power station through to the decommissioning of the reactors, including approval and licensing procedures with the regulatory government authority—is complicated.

Table 6.1 Systemic life cycle of a nuclear power station

Planning stage	Planning Application	Approx. 4 years
Construction stage	Construction operations	
Operation stage	Operation and inspection	20–40 years*
Reactor decommissioning stage	Nuclear fuel discharge System decontamination Safe storage	20–30 years* not including disposal of spent fuel
Dismantling and removal stage	Interiors Buildings	
Total no. of years		80–100 years

Source: legislation on nuclear source materials, nuclear fuel materials, and nuclear reactor regulations

*Operational life span: 60 years where extension permitted

It is also crucial to take into account the costs and time needed for the substrata inspection required before the construction of an electricity-generating station, the trial operation required before full operation, and the 'radiation-decontamination operations' necessary at the time of decommissioning, while nuclear waste in the form of spent nuclear fuel also consumes massive costs and time. The decommissioning of nuclear reactors has thus become a global issue today.

Spent nuclear fuel is stored for 3–5 years in a 'cold storage pool' within the station. Subsequent processes differ by country, but the waste is generally sent to a reprocessing plant to extract reusable uranium and plutonium, after which it is subject to long-term storage, for instance, in an underground facility at a treatment plant for highly radioactive waste. In Japan, highly radioactive waste is vitrified and kept in cold storage for 30–50 years, then disposed of underground through burial at a depth of at least 300 m in the geological strata. In November 2013, former Prime Minister Junichirō Koizumi called for an immediate end to nuclear power. To support his argument, he cited the fact that there was still no decision made on a 'spent-nuclear-fuel storage' facility and, despite the yearly increasing volume of nuclear waste, no confirmed plans as to the disposal system and technology to be used or the location of the disposal site.

6.1.2 Unstoppable Nuclear Power Generation

Today, in the wake of Japan's Fukushima nuclear accident, the world's nuclear power stations are under increasing scrutiny from the viewpoint of safety. Fukushima has taught the world that accidents could involve not only natural disasters such as earthquakes, tsunamis, typhoons, torrential rain, flooding, and drought, but also terrorism, war, coup d'état or other events that, instead of attacking the nuclear reactor itself, interrupt the functioning of the electricity-generating facilities used for cooling, causing the reactor to go into meltdown. As a result, the possibility of nuclear power facilities becoming terrorism targets has been pointed out. In France, 'Greenpeace' activists made an experimental break-in at a nuclear reactor building, while in the United States a group of three elderly protestors reportedly penetrated a nuclear reactor facility supposedly under heavy security. They are finding the 'security holes'.

From the start, the systemic life cycle of nuclear power generators, from initiation to the decommissioning of reactors, the disposal of radioactive waste, and other aspects, has remained a matter of uncertainty. As shown above in Table 6.1 (systemic life cycle of a nuclear power station), 4 years were estimated for the initiation including the initial operating period, and 20–40 years for operation, but as noted above the original 20-year life span of a nuclear power station has been extended in a common worldwide development. When decommissioning of reactors and radioactive half-life are taken into account, we arrive at a period of more than 100 years of continuing cost and labor requirements. These are not all included in calculations of the unit cost of electricity generation. There is already a history of worldwide marine disposal of drums containing radioactive nuclear waste, the cumulative total of which over 50 years has exceeded 100,000 tons according to the IAEA [5]. Figure 6.2 shows the cumulative total of sea-disposed nuclear waste by some countries.

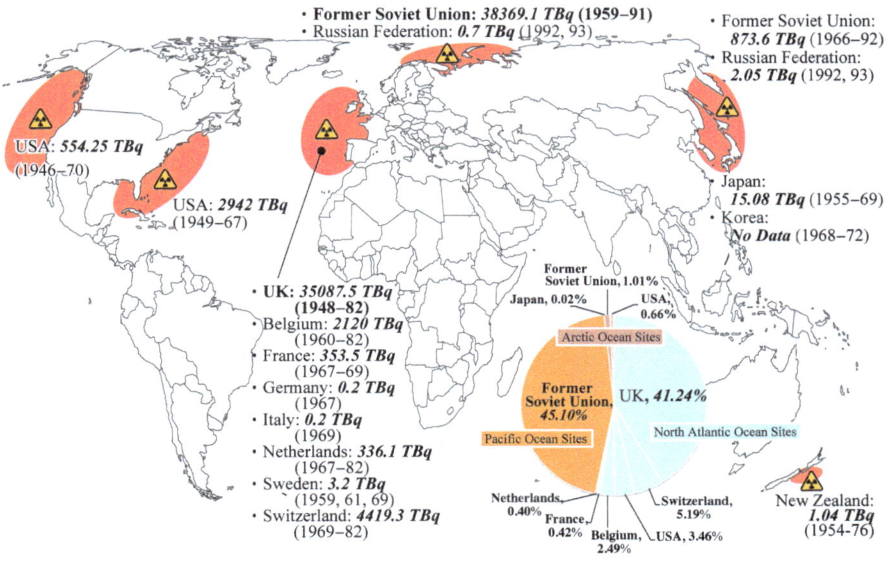

Fig. 6.2 Cumulative total of sea-disposed nuclear radioactive waste. *Note:* Mapping based on "Inventory of Radioactive Waste Disposals at Sea", IAEA-TECDOC-1105, 1999

In 1993, when an international treaty banned marine disposal of nuclear waste, America, Germany, Finland, and other countries built facilities for deep underground storage. In some cases, for instance at 'Areva's La Hague' facility in France, disposal in undersea pipelines or similar was reported. The operation of nuclear power stations thus invites Barnard's 'unsought consequences'. Nuclear power's unsought consequences or 'unexpected results' by P.F. Drucker [6] are represented in the problematic by-products of radioactive contamination from station operation: in addition to (1) limits to the manageability of nuclear power (technological issues of metal fatigue and deterioration) we also face (2) radioactive contamination, and (3) disposal of nuclear waste and decommissioning of reactors when operation ceases (legislation, systems, technology). Additionally, there is (4) the trend for local communities to petition for continued operation, for instance where local businesses have been commissioned with related projects or local governments have received legally mandated payments in return for the exploitation of electric-power resources. Thus, once a nuclear power station has begun operation, in almost all cases it continues to operate even after the inspection period is finished. This means that they are in the generation of 'unstoppable nuclear'. Once a nuclear power station is in place, it is permanent.

Underlying the "unstoppability of nuclear power generation" outlined in Fig. 6.3 are (1) a lack of standards for the decommissioning of reactor technology; (2) failure to decide on sites for disposal of accumulating nuclear waste; (3) softening up of communities local to nuclear power facilities with financial incentives through payment of compensation, consolation money, etc; (4) calculation of electricity-generation unit costs without factoring in costs for decommissioning of reactors, 'decontamination, or waste treatment'; (5) falsification of radioactive-contamination measurements; and (6) limits to the operational manageability of nuclear power stations.

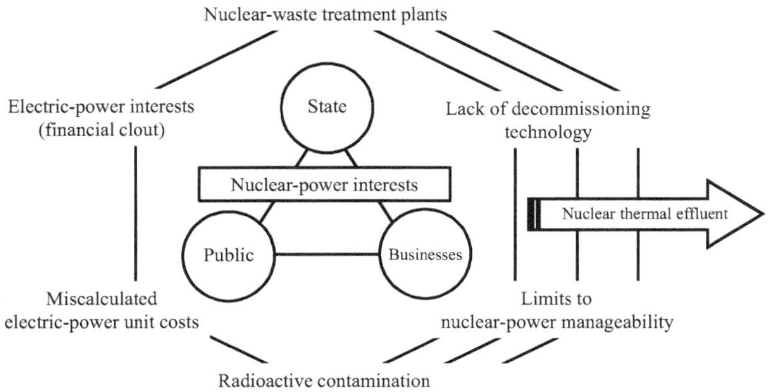

Fig. 6.3 Unstoppable nuclear power generation worldwide. *Note:* Charting based on 'Unstoppable nuclear power generation' by S. Atsuji and R. Fujimoto presented at "Systems Pathology Following the Fukushima Nuclear Catastrophe", *Social and Economic Systems Studies*, Tokyo University, 2011

Moreover, a stakeholder group has formed around the vested interests of local communities, power companies, and the regulatory 'government authorities' in charge of approval and 'licensing operations'. Massive grants from central government are not only allotted to local businesses and residents and electric-power-related associations and companies, but are also distributed in the world of academia to nuclear power research organizations and as expenses to related corporations. Among the parties involved, this is perhaps accepted to a large extent as a kind of 'tacit payment for inconvenience', 'danger money', or compensation for contamination, but as was shown by the Fukushima nuclear accident, these short-term handouts are no consolation when the worst comes to pass and the living environment, agricultural land, fishing grounds, and other resources are all lost semipermanently.

However, 'stakeholders' with connections of interest to nuclear power are not limited to the state, commercial enterprises, and local communities. The radioactive contamination that rains down on local people can cross borders to cause exposure in other areas, as at Chernobyl. It was reported by the investigation after the Chernobyl accident that radioactivity had spread across Europe [7]. In the Como region of Italy, the entire rabbit population was culled, while restrictions were placed on the export of German dairy products such as cheese and powdered milk. However, today, in Belarus and the Chernobyl district of the Ukraine, livestock and dairy farmers continue to drink contaminated milk. Of particular note is that residents of communities close to Chernobyl, and especially the children, have rates of leukemia and thyroid cancer almost five times the normal level. Following the recent Fukushima nuclear accident, radioactive contamination has been detected in coastal waters and is becoming an international issue through spread by sea currents, creating a situation for which no restitution is possible. Beyond this, what kinds of issues lie latent in regular nuclear power station effluent, which contains waste heat?

6.1.3 Nuclear Power Disasters and Radioactive Contamination (Damage to Human Health)

In the Fukushima nuclear accident precipitated by the Great East Japan Earthquake of 2011, and other nuclear power-related accidents and disasters such as the 1999 JCO criticality accident, the Chernobyl accident in the former Soviet Union, and the Three Mile Island accident in the United States, the radiation that is released in the form of cesium, strontium, and other 'elements destroys DNA', not only taking human life but also depriving people of their livelihoods. Radioactive contamination from nuclear power stations and other sources threatens human life and property and infringes on the right to life and human rights, thereby violating constitutional law; cancer and other harms resulting from radiation inflict damage on life and future generations so that it also violates criminal law; by destroying the living environment and communities and undermining livelihoods by damaging workplaces, agricultural land, fishing grounds, and other environments, it also violates civil law. The problem of radioactive contamination from nuclear power stations is a supra-legal issue not susceptible to control by current law.

Today, Fukushima's contaminated waters present a difficult problem. With Tokyo chosen as the host city of the 2020 Olympics, the International Olympic Committee views with misgiving Prime Minister Shinzō Abe's statement that "the radioactive contamination at Fukushima is completely under control." At present, as of the end of 2013, accumulated in the contaminated-water storage tanks is 334,000 tons of effluent, enough to fill 800 25-m swimming pools [8]. The cumulative total of 27,000 trillion becquerels of radioactive contamination that it has already released is said to be equivalent to approximately "1,100 times that of a Hiroshima-type atomic bomb", causing concern over the damage to local communities, human health, and the ecosystem. To illustrate the potential threat, Table 6.2 summarizes the average amount of the main nuclear species contained per ton of spent nuclear fuel, their half-life, and the potential damage to human health.

Historically, against the background of the nuclear arms race during the 'Cold War' between the United States and the Soviet Union, nuclear waste (plutonium,

Table 6.2 Half-life and damage to human body of species contained in spent nuclear fuel

Nuclear fuels nuclide	Half-life	Content per 1000 kg of concentrated spent fuel (kg)	Site of accumulation in human body/biological half-life
Uranium 238	4.48 billion years	950 kg	Bone/50 years, liver/20 years
Uranium 235	704 million years	10 kg	Bone/50 years, liver/20 years
Plutonium 239	24,000 years	10 kg	Bone/50 years, liver/20 years, reproductive glands/unknown
Strontium 90	29.1 years	26 kg	Bone/50 years
Cesium 134/137	2 years/ 30.1 years		Muscle, whole body/2–110 days
Tritium	12.3 years	–	Whole body/10–45 days
Iodine 129/131	15.7 million years/8 days	1.2 kg	Thyroid/80 days Rest of body/12 days
Americium 241	433 years	0.6 kg	Bone/50 years, liver/20 years
Neptunium 237	2.14 million years		
Curium 242	162.8 days		

Note: 1. The above table covers the main nuclear species contained in spent nuclear fuel and is not specific to the Fukushima accident, which has been confirmed to involve 31 radioactive substances
2. In Japan and other countries, after extraction at the reprocessing plant of reusable uranium and plutonium, the remaining material is buried in concrete
3. Spent nuclear fuel may be directly buried in concrete depending on the country
4. Drawing based on website of Citizens' Nuclear Information Center

californium, yellow cake, depleted uranium) from nuclear power facilities under the 'Western nuclear umbrella' was mostly collected by the United States as material for intercontinental ballistic missiles and other nuclear weapons. However, after the launch of talks under the Strategic Arms Reduction Treaty (START) between the United States and the Soviet Union, nuclear waste had to be dealt with by the individual country.

The problems associated with the aging of nuclear power stations are not limited to the decommissioning of reactors and the treatment of nuclear waste, but also involve the elevated risk of nuclear-related accidents and disasters as well as terrorism and related incidents. Accidents have already taken place at the Three Mile Island nuclear power station in the United States, [9] the Chernobyl nuclear power station in the former Soviet Union, and during the crisis accompanying the atmospheric reentry of the nuclear-reactor-equipped 'Soviet space station Mir'. In each of these cases involving the nuclear power issues of the superpowers, the facts were not sufficiently reported to other countries. However, with the collapse of the former Soviet Union, the details of the Chernobyl accident came to light, and it was reported to have caused the 'China syndrome'. This is the name for the phenomenon which occurs when a nuclear reactor goes out of control and melts and the gravitational force of the heavy uranium sends it sinking toward the center of the earth. It was found out that, as the uncontrollable nuclear reactor reached a high temperature and the building of the Chernobyl station subsided, [10] soldiers of the former Soviet army injected ultra-cooled liquid nitrogen into the ground below the reactor to prevent the reactor core from sinking. In this accident, not only were employees and local residents evacuated, but many surrounding villages and towns were also shut off. It was later found out that the radioactive substances released at Chernobyl traveled on the prevailing wind to Germany, Italy, and other nearby countries and in time to all the countries of Europe, where they spread damage by contaminating animal products.

In the JCO criticality accident of 1999, 12 years before the Fukushima nuclear accident, Japan had already experienced unexpected radiation exposure from a nuclear-power-related facility. Despite warnings of the dangers of such facilities and the systemic defects and other issues within Japan's nuclear power regulatory administration, protective measures were insufficiently stringent, and the same mistakes were repeated. It has become clear that it is no longer possible for enterprises to cope singlehandedly with the situation of a nuclear accident or disaster, which can become an issue for the government authority that decides nuclear power policy, or a focus of international conflict. This is thus a problem shared by the whole of humanity.

6.2 JCO Criticality Accident as an Organizational Disaster

6.2.1 JCO Criticality Accident Investigation: Non-risk Taking

At 10:35 a.m. on September 30, 1999, during uranium fuel-production operations in the conversion-test facility of the 'JCO Tōkai' base at Ibaragi Prefecture, a criticality accident occurred. The criticality reaction is reported to have continued for at least 20 h over a period of 2 days thereafter, during which the facility continued to emit gamma rays, neutron beams, and other forms of radioactivity. The victims of the accident were not only the workers engaged in the operation and JCO employees, but also neighboring communities. Residents living within a 350-m radius of the site were forced to take refuge for a period of around 50 h, and residents within a 10-km range were also obliged to remain indoors for a long period. Subsequently, chief cabinet secretary Hiromu Nonaka ordered evacuation of the area within a 30-km radius. After the accident, the number of people exposed to radiation, counting only those designated by the accident investigation committee, was 667 (initially reported as only 49). Among them was a group of three production-line operatives known as the 'special crew', who had been exposed to high levels of radiation, and two of whom died. Additionally, three emergency-service workers, who went to respond to the accident without being informed of its nature, were exposed to 13 mSv of radiation, while the level of exposure reached a maximum of 120 mSv among others including operatives who worked to end the criticality incident, staff at the Tōkai base, and the rescue squad. Japan's first accident at a nuclear power-related facility had claimed human lives and impacted communities neighboring the facility, becoming the worst organizational accident [11] in Japan's history of nuclear power use.

In the Tōkai criticality accident, radiation ended up escaping to the exterior as there was no concrete wall to prevent it. It had been thought that criticality would be terminated immediately after an accident in nearly all cases, but in the JCO accident, where the water coolant surrounding the settlement tank served the role of a reflective material, criticality is reported to have persisted for at least 20 h over a period of 2 days, as already mentioned. During this time, a major issue was the outward radiation of neutron beams, which have a strong ability to penetrate matter. When neutron beams collide with the 'nuclei' of the atoms making up the DNA in the body's cells, the atomic nuclei are destroyed and the DNA is damaged. Following heavy irradiation with neutron beams, cells which have suffered fatal 'DNA damage' die. The operatives who were killed were treated at Tokyo University Hospital. They were unable to regenerate their skin, and died of multiple organ failure. At the request of the bereaved families, a record of their suffering was published in a book entitled *A Slow Death*: *83 Days of Radiation Sickness* [12].

To find out whether neutron beams have penetrated the body, measurement must be made within 15 h. But initially, it was not recognized that the accident might involve the hazard of neutron irradiation, [13] and no neutron-beam measuring instrument was even available. The 667 people exposed to radiation included not only site operatives and staff at the JCO Tōkai base but also local residents. The only way to measure exposure was to estimate it from a questionnaire on the activities of the radiation victims. The government maintains that, even if cancers appear in the exposed population in the future, it will not be possible to ascertain whether these resulted from the effects of the accident. However, a health survey found that “the higher the estimated radiation dose, the greater the proportion of people complaining of symptoms” [14], suggesting that the health damage to local residents was serious. Subsequently, JCO made a uniform compensation payment to local residents of 3 million yen each, paid during the accident investigation period, the illegality of which as ‘hush money’ from an enterprise to residents was pointed out. Why, in the Fukushima nuclear disaster, was JCO’s experience of a criticality accident not drawn upon?

6.2.2 Non-crisis Management by JCO

After the criticality accident, questions were raised about the newly revealed existence of JCO’s ‘secret manual’, the state of its production and safety-control systems, and government safety inspections and regulatory administration, which were seen as causes of the accident or background factors. Why did the criticality accident occur not in a nuclear reactor, but in a facility for the production of nuclear fuel? Figure 6.4 illustrates the operations carried out by the three-person ‘special crew’ in the conversion-test facility on September 29, the day before the accident, and the operational shortcut they practiced.

As shown in Fig. 6.4, powdered Tri-uranium oct-oxide was dissolved in a special bucket and nitric acid added to create a uranyl nitrate solution. Normally, this should have been placed in a storage vat of tall and narrow shape, but before the start of operations, to ‘shorten operating’ time, the ‘labor-saving’ suggestion had been made that, instead of the time-consuming storage vat, a settlement tank should be used. This suggestion was approved by the head of the Production Planning Group, who was a qualified nuclear-fuel engineer. This change of procedure was the direct cause of the accident, but the fact that a significant change was implemented without due consideration indicates that JCO’s sense of corporate social responsibility was deficient in the area of safety and allowed risk-taking. The investigation following the accident found that this ‘illegal shortcut’ in the nuclear-fuel production process was the immediate cause of the criticality accident. The group head who authorized the modification of the operational process for labor-saving purposes testified that “the shape may be different but the volume is the same, so I thought it would be alright to use the settlement tank” [15], indicating misconceived notions of quality and quantity control and physical-shape regulation.

At the time of the accident, the simultaneous application of the procedural shortcut and the labor-saving technique precipitated criticality. The criticality conditions persisted for more than 2 days after the government’s summary

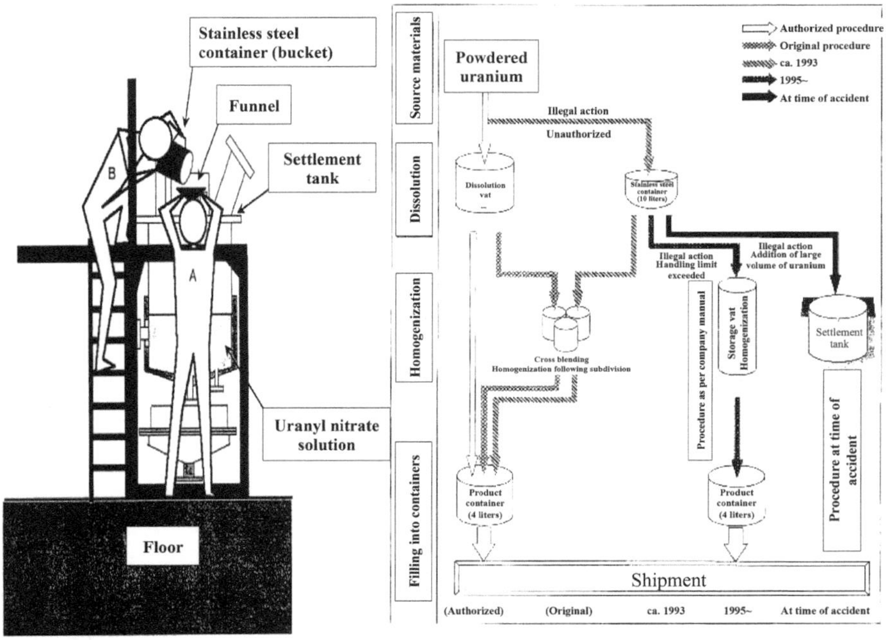

Fig. 6.4 Operations at time of JCO criticality accident and diagram of shortcut process. *Source:* Nuclear Safety Commission, Report of the Uranium Processing Plant, Criticality Accident Investigation, 1999, Figure IV-1 and Figure II-2-1

statement on the accident. Because there was no neutron-beam measuring equipment available, it is still unclear when the criticality conditions were terminated. Secondary damage was therefore caused when emergency-service workers rushed to the scene after the 'unsafe acts' [16].

As it entered the 1990s, JCO began efforts to achieve more efficient operations to accompany rationalization. First come moves to abbreviate the operational processes relating to the dissolution vat. Thus, from 1993, JCO began using a 'stainless-steel bucket' instead of a dissolution vat and switched to on-site operations based on 'human wave tactics'. This was not the operational method which had been authorized by government, and was in contravention of the quality- and quantity-control limit of one batch of uranium, designed to prevent criticality, which meant that a risk of criticality existed. However, because the storage vat was of the regulation tall and narrow shape resistant to criticality, the company was blithely confident that criticality would not occur. The factors involved in the criticality accident lay not only in the on-site operations, where shortcuts were made in the nuclear-fuel production process, but extended to production control, where the organizational laziness as *systematic soldiering* and *natural soldiering* [17] combined in an organizational accident which was 'waiting to happen' [18].

6.2.3 Systems Pathology in Japan's Nuclear Policy

What was revealed by the results of the accident investigation was that JCO had for many years neglected to carry out 'safety education relating to criticality' [19], and that production-line operatives had almost no operational experience of the production process, which meant that the special crew did not have sufficient awareness of the danger of the operation and had a low level of professional awareness as their 'career anchor' [20]. The raw materials, equipment, procedures and other items relating to nuclear-fuel manufacture are regulated in detail by the Nuclear Reactor Regulation Law, and changes to these items naturally require a government inspection and authorization. In 1996, however, with the approval of the head of the manufacturing department, JCO produced a secret in-house manual which specified illegal operations such as the use of the special stainless-steel bucket in the dissolution process and a storage vat in the homogenization process used to create uniform concentration. JCO carried on producing nuclear fuel on the basis of this manual. In the accident, this secret manual was the basis for compounding 'organizational system error' with individual human error and for bypassing regulations on container shape by decanting the uranyl nitrate solution into the 'settlement tank', which triggered the criticality accident.

Safety inspection of JCO by the government (Science and Technology Agency) was carried out under the Basic Guidelines for Nuclear Fuel Facility Safety Inspection with reference to and in accordance with the Uranium Processing Facility Safety Inspection Guidelines. JCO was bound by the provisions of Guideline no. 12: "Nuclear fuel facilities where there is a danger of criticality accidents caused by erroneous operation or other risk shall put in place appropriate measures for the eventuality of a criticality accident" [21]. However, the conversion-test facility was not equipped with a criticality alarm device to warn of criticality, or with a device for injecting neutron-absorbing material to terminate criticality; indeed, JCO did not even possess a neutron-beam measuring instrument. The safety inspection also overlooked the fact that the license application form contained mention neither of the homogenization process nor of the limit of one batch of uranium which supposedly applied to the processes for re-dissolution and homogenization. It was during this process of re-dissolution and homogenization, where the license conditions were vague, that the accident happened. As the Nuclear Reactor Regulation Law "does not stipulate compulsory regular inspection" of nuclear-fuel processing facilities, there had in fact not been a single inspection by the regulatory government authority for 'security management' [22]. Nor was there any obligation on manufacturing enterprises to report on voluntary regular inspections, and in practice the government guidance from the Science and Technology Agency, in the form of safety inspections and tours of inspection by operational-control specialists, did not constitute adequate supervision, partly because they were carried out at times when the conversion test facility was not actually in operation.

To sum up, the analysis of the criticality accident shows that it was caused by a 'system error', in other words 'organizational negligence' in the form of procedural shortcuts based on the illegal 'organizational secret manual', combined with natural negligence in the form of the human error of operational labor-saving 'self-regulation' by system error. The cause of the human error was a slipshod operational approach in which proper procedures were skipped because individual operatives wanted to make things easy and finish early. Questions were also raised over the fact that the special crew at the production frontline had not been informed of the risk of criticality in nuclear-fuel processing. The procedural shortcut, which had been devised to meet the irregularly placed orders for nuclear-fuel supply of the 'Monju Power Reactor' and Nuclear Fuel Development Corporation (PNC), deviated from the standard production process and was an egregiously dangerous form of operation. The safety inspections of the government (Science and Technology Agency) were also lax and left much at the discretion of the company. This has raised questions over the 'administrative responsibility' of central government and the Science and Technology Agency in the field of nuclear power policy, and has led to a rethink of the inspection standards of the regulatory authority. The background to the criticality accident was that insufficient consideration was given to internal and external stakeholders such as the regulatory authority, employees, and local communities, and that there was also insufficient communication within the organization about hazardous operations. The 'structural inertia' brought about by the monopolistic nature of Japan's 'fuzzy nuclear power policy' can be identified as a kind of systems pathology.

Japan's policy on nuclear power and related areas, as exemplified by the vague control of the nuclear-fuel manufacturing industry seen in the JCO criticality accident described above, is 'ambiguous'. In 2014, the foreign-affairs committee of the lower house of the Diet passed a proposed treaty on the export of nuclear power technology to Turkey, the UAE, and Vietnam under which (1) Japan would accept nuclear waste from the partner nations; (2) compensation would be provided from Japanese national taxes in the event of a nuclear accident in these nations; and (3) nuclear power operations would also be financed by Japanese national taxes. These and other conditions, favorable only to the Japanese nuclear power industry, are 'unacceptable to many Japanese citizens'. Rather than learning from the experience of Fukushima and looking toward future development of renewable energy sources, Japan runs counter to the spirit of the age by seeking to export nuclear power to Turkey, a country in an earthquake zone to which even General Electric Company and Westinghouse Electric Company do not sell. This is an unsafe nuclear power policy.

6.3 Ocean Warming through Accumulation of Thermal Effluent from the Cooling Process of Nuclear Reactors

6.3.1 Global Stakeholders

JCO used a secret manual that ordered alteration of the procedures laid down by government safety inspection standards to make shortcuts, and committed human error due to labor-saving of 'administrative systems error'. In other words it committed what F. W. Taylor calls 'systematic soldiering and natural soldiering', resulting in an organizational disaster in the form of a criticality accident. The issues involved in nuclear power accidents and disasters lead us to a moral position in which all citizens of the globe are recognized as 'various stakeholders' (A. A. Berle) [23].

In the relationship between society and the organization, there are stakeholders such as consumers, the public and shareholders, institutional investors, suppliers, partner financial institutions, the regulatory authority, and the local community. Decision-making on organizational behavior is not possible without considering these interest groups. In contrast, JCO's interest groups were limited exclusively to the public sector and included, for instance the Science and Technology Agency, which was in charge of approval and licensing operations. No consideration was shown to plant operatives and local residents through activities such as safety and 'information disclosure'. Specifically, local residents had not even been informed that the plant manufactured nuclear fuel. The risks from nuclear-fuel operations naturally do not only affect the employees, who are members of the organization, but are shared by members of the public and local residents. Corporate concern should not have been oriented exclusively toward business partners in the 'uranium nuclear-fuel supply industry' and the Science and Technology Agency, which was the regulatory authority.

The criticality accidents in nuclear-fuel-handling operations at JCO mentioned above, and the nuclear accidents at Fukushima, Chernobyl, and Three Mile Island, were not accidents limited to the locality, but disasters that grew in scale to become catastrophes that affected surrounding regions and even neighboring countries. This kind of nuclear disaster develops from an accident into a massive disaster. Thus, as shown in Fig. 6.5, the citizens of the world have felt their impact directly at a global level as 'global stakeholders'. For example, the radioactive substances released in the Chernobyl nuclear accident contaminated Europe's pasturage. Among the many examples of the associated blight was Cesium contamination of livestock, affecting cheese, milk, and other animal products in Germany and resulting in mass culling of contaminated rabbits in the Como region of Italy. Through food, water, and the air, radioactive contamination comes back to haunt humanity.

When accidents and disasters occur, the nuclear power stakeholders who end up suffering the damage are residents and members of the public. The granting of rights to nuclear power generation brought gain to some stakeholders, such as the

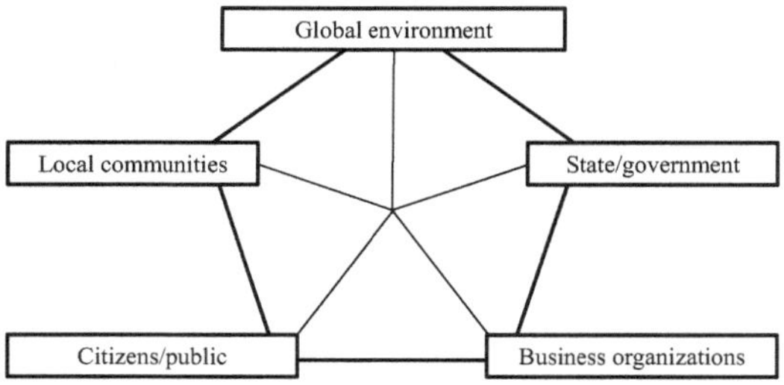

Fig. 6.5 Global stakeholders involved in nuclear power generation. *Note:* Charting based on T. Persons, H. Ulrich, J.P. Kotter, D.J. Isenberg, D.A. Gioia, and H.P. Sims

nuclear power industry, politicians close to the industry, and communities in the vicinity of power stations, but now, faced with the pressing issues of already-aging facilities and decommissioning, the negative side of nuclear-based electricity generation is becoming clear. A potential aspect of this is the contribution to global warming through rising sea temperatures due to nuclear thermal effluents from the reactor cooling process. Nuclear power electricity generation requires massive volumes of water to cool down the reactors, which reach high temperatures in the nuclear-fission process. This is why nuclear power stations in all countries are located on the coast or beside large rivers. For a nuclear reactor of average size with an output of 1 million kW, approximately 70 tons of water is required every second as nuclear coolants [24]. The waste coolant water is released into the sea or rivers at a temperature approximately 7 °C higher than when it was taken in (Fig. 6.6).

6.3.2 The Cumulative Consequence of Effluent from Nuclear Power Stations: Ocean Warming

Currently, the major cause of global warming is said to be CO_2-based atmospheric warming. However, as the specific heat of water is much higher than that of the atmosphere, the rise in sea-surface temperature due to the continuous retention of heat energy has a greater impact on global warming than the increase in CO_2. It cannot therefore be ruled out that the cumulative effect of the human-made disaster caused by 'Nuclear Coolant Effluent' is connected with natural disasters.

It is said that the rise in sea temperature—and the rise in sea-surface levels due to glacier flows which result in inflows of freshwater into the sea from the Big Melt in the polar regions, leading to lower atmospheric pressure due to the change in specific gravity, higher levels of brackish water, and further warming—will all combine synergistically to cause sea levels to rise. At COP19, the developing countries and the developed nations were at loggerheads over the question of compensation, which the former claimed for loss of territory due to sea-level rise caused by warming (November 22, 2013). Loss of territory in low-lying countries, for instance the Maldives and Tuvalu, is indeed feared due to the effect of sea temperature on global warming and rising sea-surface.

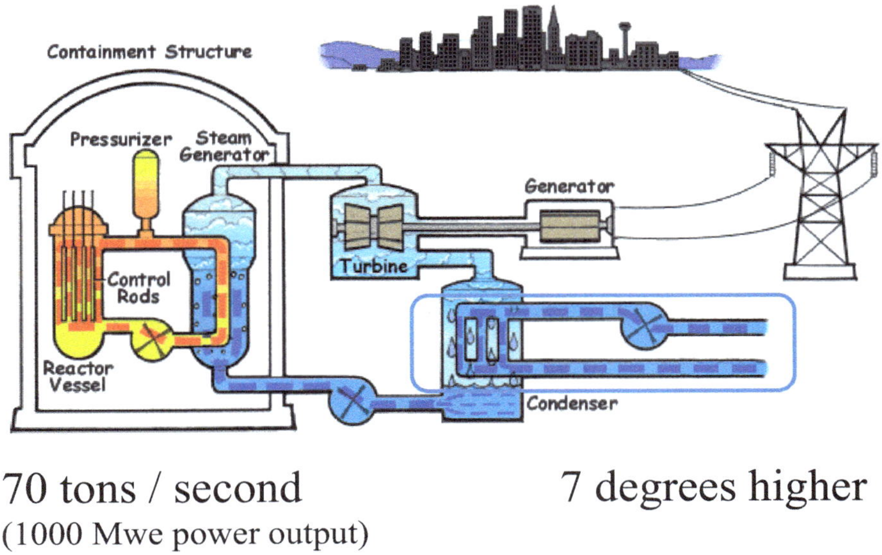

Fig. 6.6 Thermal effluents: 70 ton water/s and 7° higher for cooling process of 1000 MWe nuclear reactor. *Note:* Drawing based on United States Nuclear Regulatory Commission (http://www.nrc.gov/)

In Japan, which has 54 reactors, the world's third highest number of nuclear power stations, according to the rough calculations of Hiroaki Koide of Kyoto University's Research Reactor Institute, every year 100 billion tons of Nuclear Coolant Effluent is discharged, which means that a volume of water equal to one quarter of the total flow volume of Japan's rivers—that is, 400 billion tons—is released having been raised by approximately 7 °C in temperature [25]. Figure 6.7 shows the average percentage of operating capacity used by nuclear reactors in the main countries, indicating an operating rate of at least 70 % of capacity even as a low-end estimate after the Fukushima nuclear disaster of 2011 (previously 80 % of capacity).

Approximately 90 % of the world's population lives in the northern hemisphere, and due to the electricity consumption that accompanies economic development, 435 of the world's 441 nuclear reactors are also located in the northern hemisphere; the total of six in the southern hemisphere consists of two each in Brazil, Argentina, and South Africa. Moreover, 39.4 % of the northern hemisphere's surface area is land, compared to only 19.0 % in the southern hemisphere, meaning that the land surface is relatively large and the sea surface correspondingly smaller. Table 6.3 shows the northern hemisphere's 435 nuclear reactors by number of years of operation and estimated total volume of thermal effluent released by them, thus indicating the possibility of a 'boiling globe phenomenon' through overheating of seawater in the northern hemisphere.

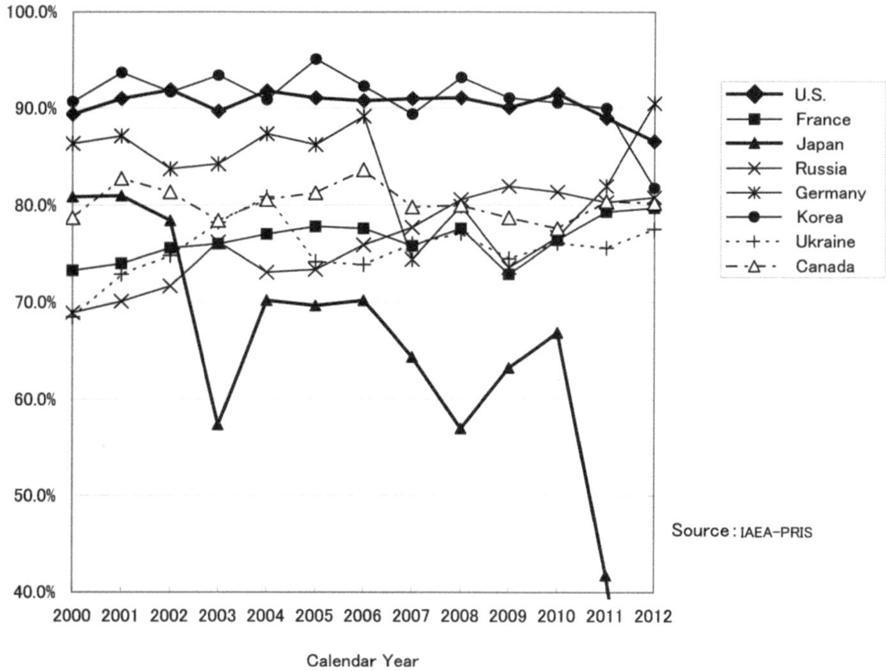

Fig. 6.7 Capacity-utilization rates of nuclear power stations in major countries 2013. *Source data:* Japan Atomic Industrial Forum, Trends in Worldwide Nuclear Power Generation, 2013

Assuming that the 435 reactors have an operating-capacity utilization rate of 70 %, have operated for approximately 31 years, and cause a 7 °C rise in 60 tons of water per second in the northern hemisphere, based on an average of 844.5 Mwe/ reactor, this means that the total discharge volume of thermal effluents amounts to 17.9 trillion tons, forming a layer of approximately 11 cm on the northern hemisphere's sea surface that consists of nuclear-based thermal effluents heated to 7 °C above the normal sea temperature or a layer of 77 cm at 1 °C above. In addition, due to convection in the atmosphere and water, temperatures are highest at the sea surface; seawater with high levels of salinity sinks, while the 'hydrosphere warming' effect is greatest on the low-salinity upper layers of water, which stay on the sea surface. The greatest influence is therefore likely to be not on sea temperature as a whole but on sea-surface temperature (Fig. 6.8). The sea covers approximately 70 % of the earth's surface, and because heat energy is more easily stored in the hydrosphere than in the atmosphere, it is retained for longer periods with long-lasting effects [26].

Table 6.3 Estimation of cumulative thermal effluents from cooling process of 441 nuclear reactors among 40 years

Area	Number of reactors	Age (Avg.)	Total power output (Mwe)	Average power output (Mwe/reactor)	Estimated total volume of thermal effluent t/s	t/year	t/40 years	t/age
World	441	30.8†	372,023†	843.6	26,042	8.21×10^{11}	3.29×10^{13}	2.53×10^{13}
Japan	55	24.8†	47,535†	864.3	3,328	1.05×10^{11}	4.20×10^{12}	2.60×10^{12}
Northern	435	30.9†	367,350†	844.5	25,715	8.11×10^{11}	3.24×10^{13}	2.51×10^{13}
Calculation	$70(\text{tons/reactor}) \times \frac{844.5(\text{Mwe})}{1000(\text{Mwe})} \approx 60(\text{tons/reactor})$ $V \approx 60 \times 864{,}000(\text{s/day}) \times 365(\text{day/year}) \times 31(\text{year}) \times 435(\text{reactor}) \times \frac{70}{100} \approx 1.79 \times 10^{13}$ $S \approx \frac{61}{100} \times \frac{1}{2} \times 4\pi \times (6.37 \times 10^6)^2 \approx 1.56 \times 10^{14}\ \text{m}^2$ $\frac{V}{S} \approx \frac{1.79 \times 10^{13}}{1.56 \times 10^{14}} \approx 0.115\ \text{m}$							
Case	Total nuclear thermal effluent of 435 reactors in northern hemisphere				Depth of sea surface occupied by thermal effluent (cm) = thermal effluent volume/sea surface area			
31 years at 70 % of operating capacity	1.79×10^{13} t				11 cm depth (7 °C rise)			

†Average number of years of operation and total power output from World Nuclear Association, Nuclear Database 2011

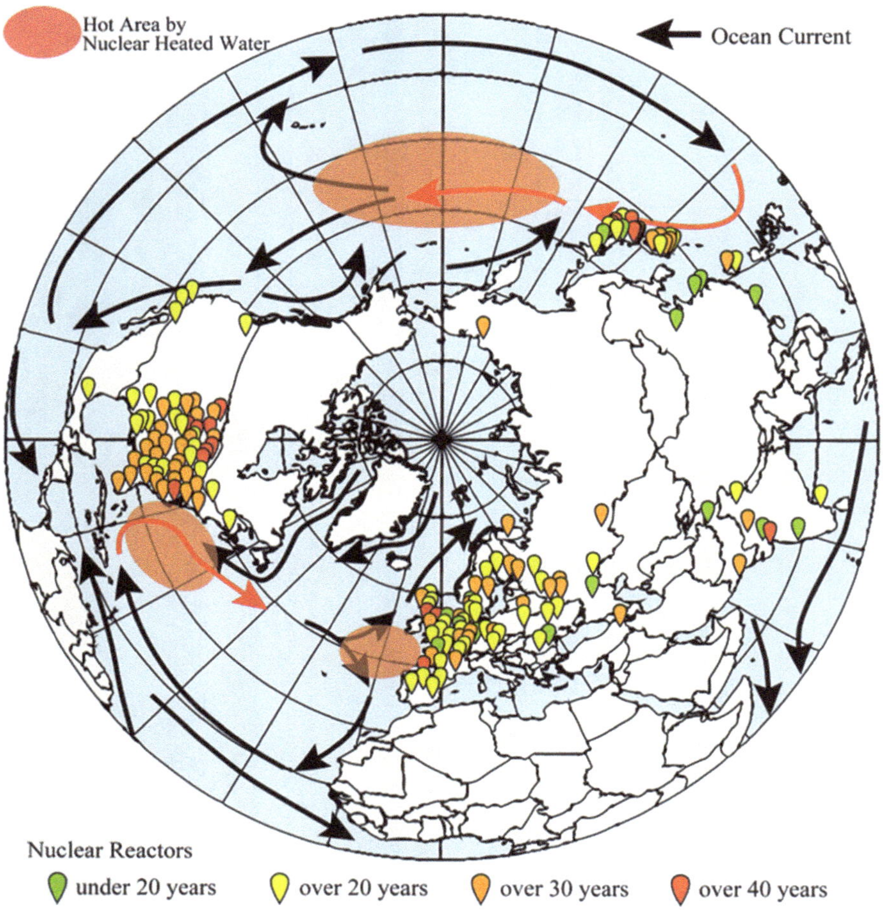

Fig. 6.8 Northern-hydrosphere warming caused by thermal effluents from cooling process of the world's 441 nuclear reactors. *Note:* Mapping by Atsuji and Fujimoto calculated using data from Nuclear Database (World Nuclear Association 2011) (with supporting calculation by Dr. N. Yoshida)

6.4 Comparison of Hydrosphere Overheating and CO_2 Atmospheric Warming

Figure 6.9 presents data on the deviation of 2013 sea-surface temperature from average values for the period 1981–2010 together with a map of the locations of nuclear power stations. As the figure shows, hydrosphere warming has risen mainly in the northern hemisphere, with a particularly high rate of risen in the North Atlantic, where there is a concentration of nuclear power stations, which may be due to the effect of thermal effluents. In a current affairs program called *Close-up Gendai* made by the Japanese broadcaster NHK, it was suggested that the rise in 'ocean warming' shown in this figure could have repercussions as far away as Asia in the form of abnormal weather patterns and natural disasters such as floods and typhoons (NHK, August 29, 2013). The program thus sounded the alarm over the potential threat of 'teleconnection'.

Approximately 2 months after the program was broadcast, on November 8, 2013, the largest typhoon ever recorded (Typhoon 30 named *Haiyan*) struck the Philippines and is estimated to have claimed over 10,000 lives, constituting a major disaster of previously unknown magnitude (NHK, November 18, 2013). The US Navy's Joint Typhoon Warning Center reported it as the strongest recorded typhoon at landfall, with maximum wind speed of 315 km/h and gusts of up to 378 km/h. Japan has also been hit by damage from unprecedented typhoons and torrential rain. In the season up to November 2013, 31 typhoons had been recorded, the first time since 1994 that the typhoon count had exceeded 30. The typhoons have also become increasingly powerful year after year, and the connection between this expanding scale and global warming has been pointed out in the Intergovernmental Panel on Climate Change (IPCC) reports [27].

In the wake of the super-typhoon that hit the Philippines (with winds averaging 324 km/h, and atmospheric pressure of 945 hectopascal), every building on the

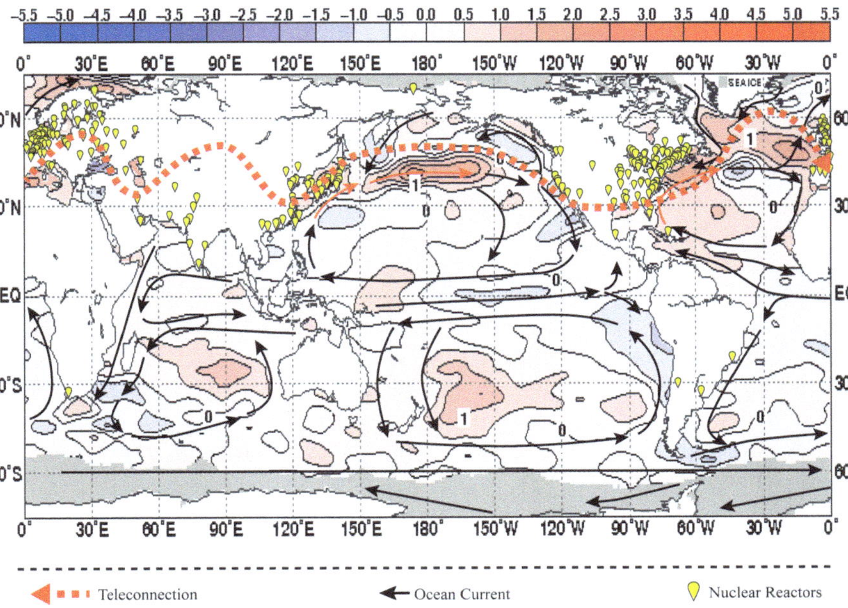

Fig. 6.9 Climate crisis from nuclear-heated oceans. Note: Redrawing of 'Teleconnection Curve' based on the map of Japan Meteorological Agency, by *permission* of JMA. *Source:* Japan Meteorological Agency (URL: http://www.jma.go.jp/)

island of Leyte was left flattened, an aftermath tragically similar to that of the Japanese earthquake and tsunami disaster of March 2011. The reach of the rainstorm's winds was limited by rising land, but the formidable winds destroyed all buildings, crops, and trees on Leyte. What caused this super-typhoon of a kind never before seen in human history? It was known already in August 2013 that the surface temperature of the sea off the Philippines had risen by 2–3 °C. Together with the atmospheric warming caused by CO_2 and methane emissions, sea-temperature rise is an insidious threat. Nuclear effluent directly heats the ocean, unlike CO_2 and methane which hold thermal energy, the possibility of 'climate crisis' by A. Gore.

The rise in sea temperature causes a problem of increased seawater volume leading to sea-level rise due to the melting not only of the land-based ice at the poles but also the glaciers of eastern Siberia, Greenland, the Arctic and Antarctica ice sheets, and the frozen seabed. The last of these causes methane gasification and further accelerates atmospheric warming. As for the buildup of nuclear coolant effluent over the years, those who have observed nuclear power generation and have seen the amounts of effluent involved at first hand will, like me, have been overwhelmed by its scale. This year-on-year accumulation of nuclear thermal effluents, amounting to 17.9 trillion tons, whose temperature has been raised by 7 °C in an 11-cm seawater layer (by 1 °C in a 77-cm layer), cannot be ruled out as a factor in the abnormal weather patterns observed worldwide.

The IPCC also predicts a rise in atmospheric temperature of up to 4.8 °C by 2100, which will be accompanied by an 82-cm rise in sea levels (Fig. 6.10), while the 19th session of the Conference of the Parties to the United Nations Framework Agreement on Climate Change (COP19) in Warsaw, Poland, saw conflict as developing countries insisted that developed nations should compensate them for floods and typhoons caused by warming and for loss of submerged territory and other damage. This clash between developed and developing nations led to an impasse at the 2013 conference, which had to be extended. The extreme phenomena already being reported worldwide include unprecedented super-typhoons with air pressure under 900 hectopascal, great floods, summer snowfall in France, floods in Germany and Austria, landslides in Vietnam and Japan's Izu Islands, and massive tornados in America. Posited as a remote cause of this is the possibility of 'global ocean warming' caused by effluent from the nuclear power plants? If true, this demonstrates that "a chain of human-made disasters adds up to natural disaster" and that ultimately the two types of disaster are intertwined via teleconnection (*see also* Fig. 5.6: The title of Fig. 5.6 is actually Global distribution of countries or areas at risk of Dengue transmission).

In a project to predict the situation with global warming around the end of the 21st century, conducted by research groups including the Meteorological Research Institute of the Japan Meteorological Agency and the Advanced Earth Science and Technology Organization [28], it is stated that the number of very strong tropical low-pressure systems with maximum wind speeds over sea or land of more than 162 km/h is on a rising trend due to the increasing scale worldwide of typhoons, hurricanes, and cyclones, with a rise in sea level caused by extremely low air

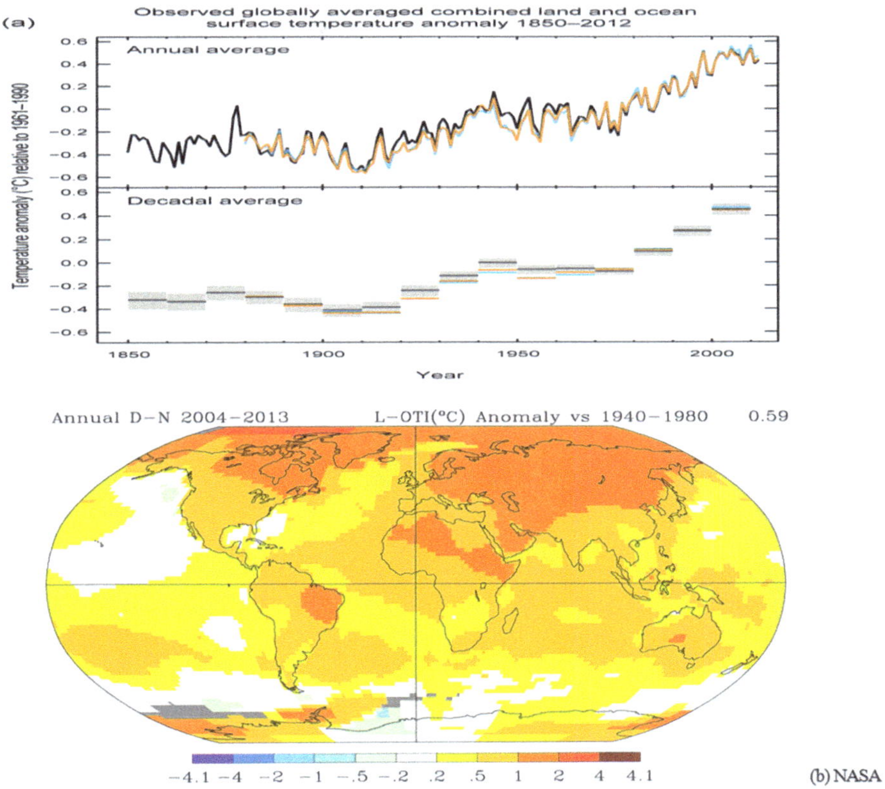

Fig. 6.10 Global warming by IPCC and NASA. *Source:* (**a**) IPCC 2014 and (**b**) NASA 2014, *permitted* by website policy

pressure like the 895 hectopascal of the Haiyan super-typhoon of 2013, and other weather events accompanying global warming. If human activity, including such Pacific Ocean warming, is promoting global warming, then there is, underlying natural disasters such as typhoons and torrential rain, floods and earthquakes, an accumulation of human-made disasters. As with the Fukushima nuclear disaster, which arose out of the Great East Japan Earthquake, there are cases where natural disasters develop into human-made disasters; but conversely, there are also cases where natural disasters develop out of unnatural disasters. The cumulative chain of human cooperation thus creates a situation where "human-made disasters and natural disasters are interconnected", and contributes to unsafety.

Even more so than the global warming caused by the effect of CO_2 and methane on the thermal energy of the globe, the boiling globe effect of the total thermal energy of heated water has the potential to produce 'unexpected consequences'. For instance, the permanently frozen glaciers of Greenland and Antarctica could melt under the influence of global warming and form a moraine. The moraine would cause water to flow in and gather 2–3 km below the ice sheet, which would lift the glacier so that the whole ice cap might plunge into the sea in one piece, setting off a huge wave. As the ice cap lifted with further melting, the rising sea-levels of the continents of Greenland, Siberia, the Arctic and Antarctica, which had until then been subject to subsidence of several kilometers, might suddenly be forced upward and release a tsunami. In such an event, there would be a threat of un-safety to the coastlines of countries around the North Sea, such as Norway, Iceland, and Great Britain.

The possibility of global warming from the boiling globe phenomenon caused by the accumulation over 40 years of thermal effluent by cooling process from the world's 441 nuclear reactors, which are presented here as examples, points precisely to the "interconnection of human-related natural disasters". This highlights further aspects of the multiple impacts of nuclear power on the environment, including atmospheric and marine pollution in the event of an accident, and nuclear effluents. The impact on the environment means direct impact on the ecosystem and people living in the environment. Going forward, in response to the danger which has passed the limits of manageability by businesses and central governments, there is a need for a preventive social function to oppose the collusive relationship between business management and government policy over nuclear power. Essential here is a global eco-civilization in which individual citizenship has preventive power.

References

1. Atsuji, "Un–safety: Systems Pathology of the Fukushima Nuclear Catastrophe", *ISSS Proceedings*, 2013 and Atsuji et al., "Sustainable Decision–making Following the Fukushima Nuclear Catastrophe", *IFSAM*, 2012. Presented at ISSS 2013 (Vietnam), IFSAM 2014 (Tokyo) and ISSS 2015 (Berlin).
2. Tōkyō Denryoku Fukushima Genshiryoku Hatsudensho Jiko Chōsa Iinkai [Tokyo Electric Fukushima Nuclear Plant Accident Commission], Kokkai jiko chōhōkokusho [Accident Report to the Diet], Tokuma Shoten, 2012.
3. Yablokov, A.V., Nesterenko, V.B., and Nesterenko, A.V., *Chernobyl: Consequences of the Catastrophe for People and the Environment*, Blackwell, 2009.
4. Agency for Natural Resources and Energy, Genshiryokuhatsuden no haishisochi wo meguru kaikeiseido no kadai to ronten [Issues and Questions Concerning Accounting Systems for the Measures around Nuclear power Decommissioning], 2013.
5. IAEA, "Inventory of Radioactive Waste Disposals at Sea", IAEA–TECDOC–1105, 1999.
6. Drucker, P. F., *Innovation and Entrepreneurship*, Harper Collins Publishers, 1985.
7. Yablokov, A.V., Nesterenko, V.B., and Nesterenko, A.V., *op. cit.*
8. *See for example,* Tokyo Shimbun, 23 August 2013.
9. Kemeny, J., *The Need for Change: The Legacy of TMI. Report of the Presidential Commission at Three Mile Island*, New York: Pergamon, 1979.
10. Reason, J., "The Chernobyl Errors", *Bulletin of the British Psychological Society*, Vol. 40, 1987, pp. 2201–6.
11. Reason, J., *Managing the Risks of Organizational Accidents*, Ashgate Publishing, 1997, pp. 1–20.

12. NHK Tōkaimura Rinkai Jiko Chōsahan [Tokaimura Criticality Accident Investigation], Kuchiteitta inochi—Hibaku chiryō 83 nichikan no kiroku, Shinchōsha, 2006.
13. Awano, M., "Ano hi Tōkaimura de nani ga okotta ka" [What Happened That Day at Tokaimura?], Nanatsumori Syokan, 2001, p. 103.
14. *See for example,* Mainichi Shimbun, 23 September 2001.
15. Awano, M., *op. cit.*, 2001, p. 90.
16. Reason, J., *The Human Contribution: Unsafe Acts, Accidents, and Heroic Recoveries,* Ashgate Publishing, 2008.
17. Taylor, F. W., "Shop Management" (in *Scientific Management*, 1947), Harper & Brothers, 1903, pp. 30–34.
18. Reason, J., *op. cit.* (*Managing the Risks of Organizational Accidents*), 1997.
19. JCO Rinkai Jiko Sōgō Hyouka Kaigi [Meeting Presenting Overall Assessments of the JCO Criticality Accident], *JCO rinkai jiko to Japan no genshiryoku gyōsei [The JCO Criticality Accident and Japan's Nuclear power Administration]*, Nanatsumori Shokan, 2000, p. 35.
20. Schein, E.H., *Career Dynamics: Matching Individual and Organizational Needs*, Addison–Wesley Longman, Incorporated, 1978, pp. 171–172.
21. *See,* Nuclear Safety Commission, "Kakunenryō shisetsu anzen shinsa kihon hōshin" [Basic Measures for Safety Inspections of Nuclear–fuel Facilities] (http://nsc.jst.go.jp/anzen/sisin/sisin025/si025.html).
22. Nuclear Safety Commission, Uranium Kakō Kōjō Rinkai Jiko Chōsa Iinkai [Uranium-Processing Factory Criticality Accident Investigation Committee], "Uranium Kakō Kōjō Rinkaijiko Chōsa Iinkai Hōkoku", [Uranium-Processing Plant Criticality Accident Investigation Report] 1999, III–1–(6).
23. Berle, A.A., and Means, G.C., *The Modern Corporation and Private Property*, Transaction Publishers, 1932.
24. Koide, H., *Genpatsu no uso [The Lies about Nuclear Power]*, Fusōsha, 2011, p. 119.
25. Koide, H., *Ibid.*, p. 120.
26. S. Atsuji, et.al., "Disaster Management for Boiling Globe by World's 441 Nuclear-heated Water", *Program & Abstracts 2015,* (ISSS Journals), pp.134.
27. IPCC: "Intergovernmental Panel on Climate Change", *Climate Change 2013: The Physical Science Basis*, 2013.
28. Meteorological Research Institute of the Japan Meteorological Agency and the Advanced Earth Science and Technology Organization.

Part III
Science of Crises

Itsuki goes on to say that words that touch the heart and which are 'repaid in emotion' are information in the truest sense. The true significance of this information is surely that it is repaid by emotion promoting empathy, coexistence, and symbioses. Information as such consists not only of communication through the symbolic medium of letters and figures, but also of an interconnection between human minds including the methods by which it is transmitted, and encompasses the communication of S.K. Langer's 'import' in which human emotion is directly expressed. Communication will not take place unless it is rooted in an empathy which is in tune with the range of human sensibility common to all individuals, which is also the source of human relationships of mutual trust. It is the human intelligence that underlies the reconciliation and integration of reason and sensibility that is required to realize this 'emotion and reason that touch the heart and soul'. This is precisely the human 'intellect, emotion, and volition' that give rise to goodness and virtue, and well-being for human kind.

Chapter 7
Escape from Disaster: Invisible Informatics of Risks and Crises

Why are the lands that were home to four of the world's great civilizations—Egypt, Mesopotamia, the Indus valley, and China—turning to desert? What has gone wrong? Is it human-made disaster or natural disaster? The Maya and Inca civilizations, Easter Island, and many other ancient civilizations collapsed and vanished together with the forests [1]. It is said that Greece and the islands around the Aegean Sea were originally covered in forest, which was used up as building material and fuel, causing desertification. In Africa, too, whence humanity originated, the progress of desertification has brought water shortages and drought conditions that have plunged the continent into food crisis.

Historically, in the ancient civilizations of Egypt and Mesopotamia, the Indus valley, and China, because river floods would damage the crop harvest, it was recognized that 'to rule the waters is to rule the land'. Use of astronomy to forecast floods and drought can be said to have given rise to the power of the state [2]. In the history of humanity, the ability to predict disaster was the basis for the state's existence. However, the Fukushima nuclear disaster revealed a tendency for the state to falsify and distort information about the disaster before releasing it to the general population. For the sake of maintaining state power, and with the aim of downplaying the impact of government negligence, the responsibility of the relevant departments, and the fiscal burden of disaster recovery, there is now frequent manipulation of information fed through the media to victims, local communities, and the Japanese public in general.

This trend resembles the wartime information manipulation under Japan's Imperial government, when the Imperial General Headquarters would report that battles actually lost had been won. This is a phenomenon not limited to Japan but seen in many states. The falsification and concealment of information and the manipulation of information by the state power are on an escalating trend and have reached a level where a revision of the role of the state in disaster situations is needed. In all ages and all parts of the globe, the unfortunate events of war, disaster, and accident have tended to be forgotten. As a result, they have been repeated historically time and again, and it is difficult to say that their lessons have been sufficiently learned.

S. Atsuji, *Unsafety*, Translational Systems Sciences 7,
DOI 10.1007/978-4-431-55924-5_7

If one studies documents recording war, disaster, and accidents from whatever part of the world, the more one learns of the facts, the more they seem stranger than fiction. Torahiko Terada warned, "disasters turn up just when we have forgotten the last one" as 'biological weakness of human nature' by Erich Fromm [3].

The present chapter firstly examines risk and crisis, concentrating on the invisible information process of unsafety; secondly it addresses the mechanism of 'pseudo escape' despite cognition of the signs of disaster; thirdly, it examines the circumstances whereby natural and unnatural disaster can combine to precipitate an unforeseen human catastrophe [4]; and fourthly, it considers the possibility of 'disaster management'. The chapter will also present sustainability-related initiatives from overseas that support the coexistence of water- and carbon-based life on Gaia. How to limit unsafety is not a question I can address to the people of the future, or to anyone but ourselves. Someone must therefore speak up for the global environment. But is it to be politicians or businesspeople or instead the citizenry?

7.1 Risks and Crises of Hidden Unsafety

The meteorite which suddenly crashed down on Siberia in 2013 was an example of the way the earth's climate is influenced by phenomena which humankind is powerless to affect, such as cosmic radiation from supernova explosions in the galaxy and solar activity. The alignment of planets which occurred in the solar system around 2012 caused an imbalance in the forces of gravity, which in turn affected geological mantle convection and plate movements, boosting volcanic and seismic activity and triggering earthquakes and tsunamis, extremes of cold and heat, floods, and avalanches. The succession of ice ages and other cycles in the 'climate change' bring alternation between the 'snowball earth' when the whole globe is frozen, and 'big melts' when all the ice is dissolved [5]. There are global phenomena which humankind cannot escape.

Chemical contamination caused by environmental hormones and atmospheric pollution (Chap. 3) and global warming caused by CO_2 emissions and methane vaporization lead in a chain from acid rain, forest destruction, and desertification through floods, water shortage, and drought to crop failure, food crisis, and hunger and famine. Coinciding with the explosion in the world's population to 9 billion in 2050 by J. Cribb, abnormal weather patterns are exacerbating the food crisis caused by crop failure [6]. Due to 'global warming', infectious tropical diseases and endemic diseases are spreading north and south, disseminating pathogens worldwide (Chap. 5), so that the compounding of natural disasters with human-made disasters results in complex catastrophes. The 'boiling hydrosphere' caused by thermal accumulation among 40 years of effluents from the cooling process of the world's 441 nuclear reactors has resulted in ocean warming (Chap. 6), which contributes to the potential for abnormal weather patterns caused by atmospheric CO_2-based global warming. The human-made 'wealth imbalance' between north

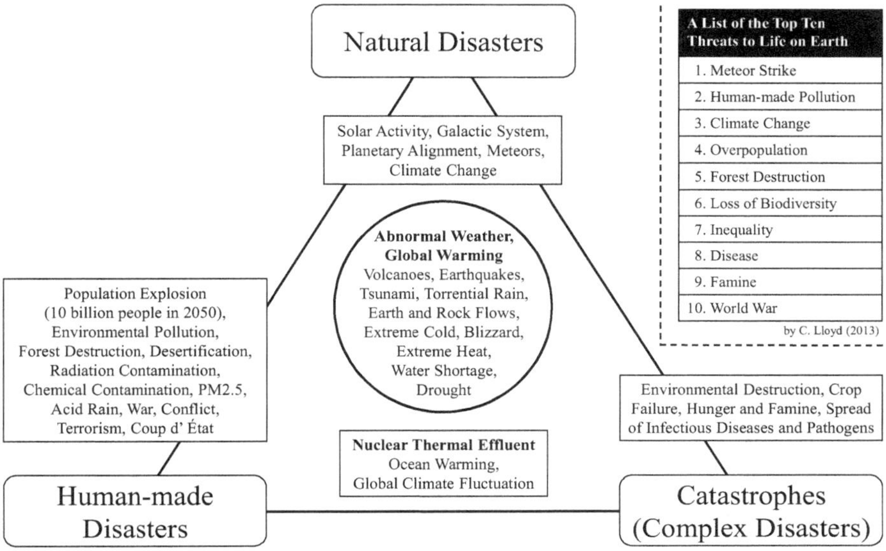

Fig. 7.1 Unsafety triangle: interrelation between natural and human-made disaster. *Note:* Drawing based on *World on the Edge* by L. Brown and *What on Earth Happened?* by C. Lloyd

and south wealth, the clash between religions, and ethnic issues precipitate war, conflict and terrorism, and are further wellsprings of *unsafety*.

C. Lloyd's *What on Earth Happened* contains a list of the top ten threats to life on earth: (1) meteor strike, (2) human-made pollution, (3) climate change, (4) overpopulation, (5) forest destruction, (6) loss of biodiversity, (7) inequality (poverty, military action, terrorism), (8) disease, (9) famine, and (10) world war [7]. In each scenario, it is predicted that natural disasters will be compounded with unnatural disasters to create complex catastrophes which threaten life on earth (including humankind). Here, the author has combined Lloyd's predictions with the content of Parts I and II of this book (particularly Chap. 6: Boiling Globe) to create an illustration I have called the 'unsafety triangle' (Fig. 7.1).

The reasons why disasters and accidents are forgotten and then repeated are not only that there is a limit to linguistic recognition in information cognition by the individual or public, but also that there is a basic problem in the information processing of social organizations and systems. In the great tsunami that followed the Great East Japan Earthquake, the people who managed to escape tended not to be those who complied with the disaster-response measures issued by government institutions, but rather those who, in keeping with age-old tradition in times of tsunami, headed straight for high ground away from the coast, where monuments to past tsunamis stood (NHK, September 9, 2013). When we hear about disasters and accidents, apart from when we are involved ourselves, we see and hear from the victims and their acquaintances and relatives, and also gather information through the media, and base our understanding on an image we create. When we seek to

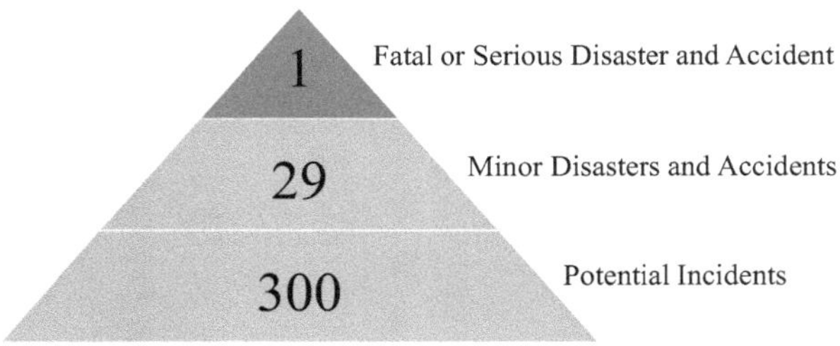

Fig. 7.2 Incident pyramid by Heinrich's estimation. *Note:* Charting a suggestion by G. Chroust based on H.W. Heinrich, *Industrial Accident Prevention – A Scientific Approach* [8]. *See also* 'Disaster Management' by S. Bennett

understand disasters and accidents without direct experiences, we do so first of all through cognition of this self-created image, which we then set against analogous past experiences, thus establishing a kind of pseudo-awareness. The information process of learning about a disaster or accident does not lead to adequate understanding if it cannot be based in the human brain as the biological memory of something already experienced. The information process in this recognition of disasters can be seen as mediated through an invisible process.

Looking back over previous research and classics in the 'magnetic field of unsafety', one finds Heinrich, in 1929 at the time of the Great Depression in the United States, setting out a 'pyramid model of disasters' with regard to human unsafety behavior (Fig. 7.2).

He addresses chiefly the subject of the sharp increase in work disasters during America's period of high economic growth, and demonstrates that a single serious accident is mediated through 29 separate minor accidents, which in turn are founded on around 300 latent irregularities or 'near-miss' potential incidents [8]. Heinrich's approach shows that behind accidents and disasters there is a series of prior stages. As for signal perception or image cognition in advance of these disasters and accidents, they cannot be perceived with the five senses alone, and include unknowable information which we can only access through previous generations who have already experienced them. From signals which can be perceived with the five senses, we cognize an advance signal, which leads to an advance image, the transmitting of which is the process that forms the stage prior to linguistic recognition. Advance signal perception in disasters and advance image cognition in accidents can be seen as signs for recognizing unsafety [9]. The process by which we learn of disasters and accidents is different depending on whether we experience them directly or linguistically. We objectify disaster phenomena and understand them as abstracts by encoding them in language and figures. However, knowing of past disasters means no more than symbolifying and abstracting certain aspects, with much of the rest of the information discarded (Fig. 7.3).

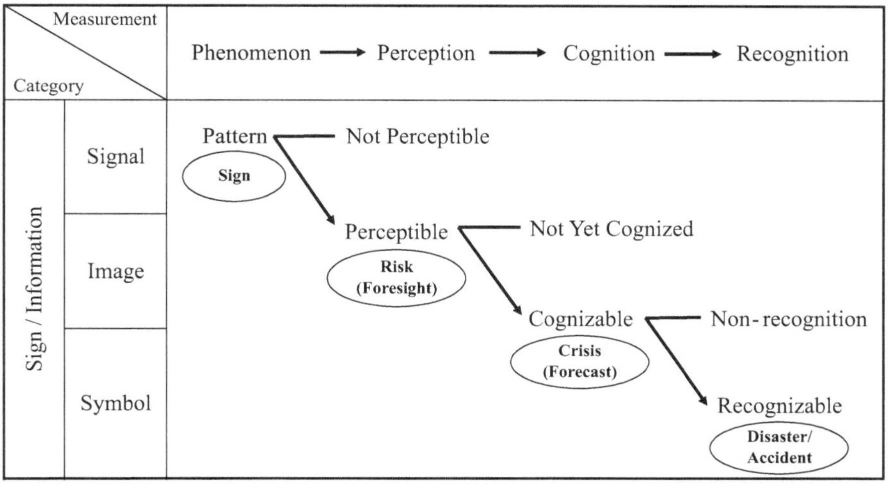

Fig. 7.3 Tacit dimensions of unsafety (risk-crisis-accident-disaster-catastrophe). *Note: Informatics* concept from G. A. Miller, H. Gardner, J. R. Pierce, C. Eccles, H. I. Brown, F. Capra, W. R. Ashby, and D. Bohm. *Source:* 'information' based on *Cybernetics* by N. Wiener with 'import' on *Problems of Art* by S. K. Langer)

7.1.1 Signal Perception (Advance Perception of Disaster Signs)

Of the signals perceived by the five human senses [10], some such as light and sound, air and other elements of climate, and astronomic signs are perceived without particular awareness. The rumblings of the earth, the movements of clouds in the atmosphere, and the stirrings of the sea are among the signs that can evoke advance 'nonconscious' signal perception of disasters, but only people with previous experience and a small number of experts will notice them, and many of them are unsuited to ready social transmission.

7.1.2 Image Cognition (Advance Image Cognition of Risk or Crisis)

The perceived signals are cognized in a disjointed fashion, vaguely, and without conscious recognition [11]. Risk or crisis sometimes takes on the semi-physical form of an image, but is latent and not easily objectivized. It is in the state described by M. Polanyi as intuitive indwelling or preconscious awareness [12], and remains in the 'tacit dimension' of the invisible process. There is one seismologist who predicted the Great East Japan Earthquake. He studied the geological stratum of the Tohoku region's coastal strip, researched historical texts from before the Edo period, identified the cyclic pattern and the relevant area, and sounded the alarm (Tōhoku University). Neither the state nor the government nor academic circles responded to his prediction, but it turned out to be accurate. Prediction of risk and crisis based on the historical records of our ancestors was thus more accurate than the latest measuring instruments and had stronger predictive ability than the most up-to-date seismology.

7.1.3 Recognition (Recognition of Disasters and Accidents)

From signals perceived through the five senses, we cognize an image and encode it as letters and numerical value, which allows linguistic understanding, communication and thought. Signs of disaster are encoded from the image, and in this process of giving form, it becomes the object of language-based thought. When we discern the risk of disaster or a crisis, we communicate verbally, which makes us aware of the associated unsafety. In connection with this invisible information process, H. A. Simon posits the concept of a 'focus of attention' [13] and C. I. Barnard the 'zone of indifference' [14] as a way of clarifying the tacit information process between the individual and societal levels. Signs of disasters and accidents are perceived as intangible signals, which develop into cognition of crisis or risk, and are verbalized to become the object of thought and communication in what is termed the implicit process of knowledge (Fig. 7.3, above). As shown in Fig. 7.4, signs of disasters and accidents progress from the signal stage (perception) through an image stage (cognition) to linguistic recognition, in which processing a large amount of abstraction (discarding of inessential information) takes place and a tacit dimension intervenes.

The so-called well-structured level of H. A. Simon's hierarchy consists of encodable alphanumeric data within linguistic knowledge, while the quasi-structured level is a nonlinguistic domain, including emotion, which mediates behavioral knowledge and sensorimotor bodily knowledge. Additionally, the ill-structured level defines 'value premises and factual premises' based on the human 'bounded rationality' of H. A. Simon [15]. Although value premises and factual premises are normally not consciously recognized, these are formed by a tacit information process in the domain of cognition. Nonsymbolic image processing and symbol manipulation is the essential part of human intelligence which is processed by an invisible informatics category, but is limited by the focus of attention. Individual information cognition is stimulated by interaction with the mutuality of social relations and is cultivated by an invisible interaction between the individual and the collective involving the intelligence of the collective and the intelligence of the individual. Accordingly, when seen in connection with the multilayered information structure involved in advance signal perception of the signs of disaster and accident and the cognition of risk and crisis, as outlined above, the following becomes clear: as shown in Fig. 7.4, the hierarchy in which signs of disaster and accident are perceived, and then cognized, and reach the level of explicit

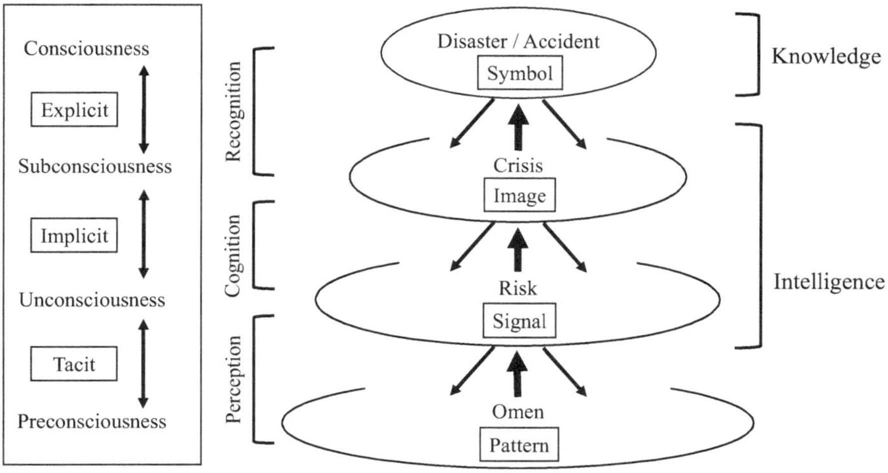

Fig. 7.4 Hierarchy of implicit cognition in personal unsafety. *Note:* Drawing based on the consideration from "World Organization of Systems and Cybernetics" by S. Beer presented at ISSS 2000. *Source:* 'Cognitive Science' LNR group of P. Lindsay, D. Norman, U. Neisser, and D. Rumelhart

and implicit awareness involves a sequential progression through human brain preconsciousness, unconsciousness, latent subconsciousness, and consciousness, such that a tacit intellectual process mediates in the emergence of conscious recognition [16].

Regarding latent human intellectual activity, M. Polanyi describes it as 'tacit knowledge' [17]. The English philosopher A. J. Ayer identified the empirical knowledge which is gained through activity [18], while C. I. Barnard introduces the role of 'behavioral knowledge and bodily knowledge' with reference to G. Harbord and Roberto Michels [19]. Many scholars have focused on these latent areas of human intelligence. Unsafety, in connection with invisible tacit knowledge, has allowed the signs of disaster and accident to be overlooked, based on vulnerable systems [20]. Tacit knowledge can be said to mediate within the invisible latent areas of humans and society.

7.2 Tacit Zone of Indifference: Accident and Disaster Signs

The Fukushima Nuclear Power Catastrophe (Chap. 2) associated with the Great East Japan Earthquake of 2011 and the JR Fukuchiyama Line derailment accident of 2005 (Chap. 4) [21], caused great anguish, so why do people tend to forget such things? The same question can be asked at the United Nations about the Afghan War, the Gulf War, and the Iraq War, hurricanes Katrina and Andrew, or the Saint Louis floods. In Europe analogous examples include volcanic activity in Iceland, major floods, and extreme heat. Such incidents leave a strong impression on people's consciousness at the time, but at some point, their consciousness alters, leading to repetitions of the same disasters and accidents. The disasters people experience naturally differ depending on the country and the region, ranging from earthquakes, tsunamis, typhoons, floods, volcanoes, avalanches, earth and rock flows, and geological fissures, to terrorism and war.

The US city of San Francisco has suffered major earthquakes in the past. As we know, the Bay Bridge was destroyed by one in 1985. Having experienced earthquake damage, the city emphasizes quake resistance and disaster preparedness in urban planning for its buildings, subway, bridges, and highways. These initiatives have a lesson to teach Japan, with its frequent earthquakes. Japan's ancient wooden structures dating back as far as 1300 years, foremost among them being the pagoda of Hōryūji Temple, have withstood even major earthquakes. Why is this, when comparatively recent buildings have collapsed?

Not all the disaster scenarios described in Table 7.1 would be comprehensible to people from other countries. People and countries that have experienced the same kinds of disaster are able to appreciate their horror and tragedy. On the other hand, there are in fact people who, despite having experienced the menace and fear of a disaster, try to forget it in order to escape from the trauma. There is a Japanese saying: "Once it's swallowed, you forget how hot the food was." People are afraid to remember. If unpleasant past episodes are constantly kept fresh in the consciousness and continue to be the subject of concern, the resulting prolonged tension affects one's personality. Despite swinging between animation and depression, by building new memories, people forget the fear that stalks them and unconsciously

Table 7.1 Categories of disaster level by cognitive domain in crisis experience

Intensity level	Earthquake intensity, volcanic eruption	Wind speed	Typhoons, tropical low pressure system	Rainfall	Heavy rain, floods, torrential rainstorms
1	Slight shock	10 m/s	Raindrops painful	10 mm/h	Below-floor flooding
2	Strong jolt	20 m/s	Unable to breathe	20 mm/h	Above-floor flooding
3	Unsteadiness requiring support with the hands	30 m/s	Body lifted off ground	30 mm/h	Road flooding, transit systems suspended
4	Objects fall	40 m/s	Unable to stand	50 mm/h	Mudslides, earth and rock flows
5	Unable to stand	50 m/s	Objects fly around	80 mm/h	Landslides, surface landslips
6	Objects fly	60 m/s	Cars and yachts fly through air	120 mm/h	Like a tsunami
7	People fall over, crouch down	70 m/s	Trees and buildings collapse	150 mm/h	Deep landslides

Note: Classification of 'Experienced information' based on P. L. Berger, T. Luckman, J. E. Hochberg, and W. R. Reitman seek mental tranquility. It is an emotional strain to remain constantly in the 'zone of concern' in response to the unsafety of the disasters that strike in a cyclical pattern. Instead, in the face of the misfortunes which may strike whenever and wherever, and inescapable disaster, one erects a temporary façade of indifference in order to carry on with everyday life. To deal with earthquakes and typhoons which strike without warning, or terrorism and accidents, one has to tell oneself that one will be all right or one could not carry on as normal.

As shown in Fig. 7.5 below, at certain times, accidents and disasters are perceived as objects of people's concern, but with the passing of time, the zone of concern contracts and moves on. People's zone of concern is not sustained permanently but constantly changes as one's powers of attention are dispersed with the emergence of new events. C.I. Barnard uses the term 'zone of indifference' [22], which the present author applies to the current issues of risk and crisis.

The origin of unsafety is in human material greed and inattention. The habit of assimilating to a fiction of safety and peace of mind in order to escape from risk and crisis is no doubt a feature of people's deep psychology common to all humankind. The range of people's power of attention is limited, and, under pressure from

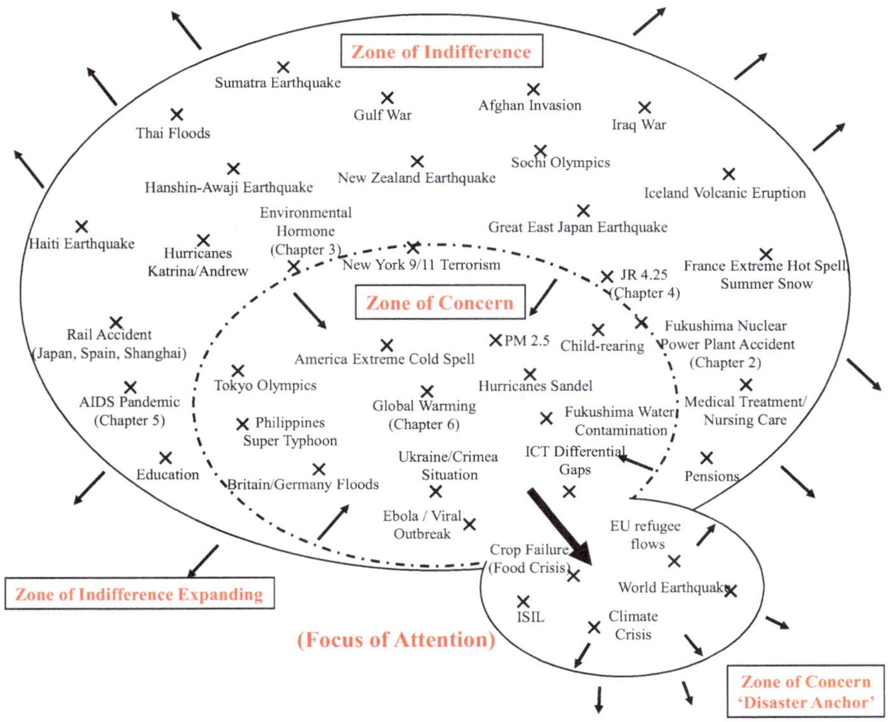

Fig. 7.5 Cognitive domain of 'disaster anchor' based on crisis experiences. *Note:* Drawing based on integration from 'zone of indifference' by C. I. Barnard and 'focus of attention' by H. A. Simon

various factors, concern with risk and crisis moves on, powers of attention are dispersed, and the zone of indifference expands.

In Japan, following the awarding of the Tokyo Olympics by the 125th IOC Session, there is currently a proposal to locate one of the venues in Fukushima Prefecture (February 2014, *Mainichi Shimbun*). This may have been no more than a suggestion, but it would be rather like setting up a venue for the Sochi Olympics at nearby Chernobyl. Whichever the event involved, one would have to feel sorry for the athletes. This proposal, made by government officials, was completely contrary to the spirit of the Games catchphrase 'omotenashi', hospitality. As underlined by the phrase "what's common sense in Japan is nonsense in the rest of the world", there are still many Japanese who only follow domestic affairs and have no sense of internationalism. The Japanese zone of concern has become paralyzed because of the excessive number of disasters. "Quick to heat up, quick to cool down", this forgetful nation has perhaps been worn down by repeated disaster and left indifferent. The issue of radiation-contaminated water from the Fukushima nuclear power station may also be explained by the national character of a people who find that "once swallowed, food is no longer hot". In Japan, there is a tendency, criticized by the overseas media, to use the argument that "no one died at the nuclear facility" to neutralize public concern in favor of the continued operation of nuclear power facilities. But in the JCO criticality accident, the victims of radiation included the local residents, the JCO's special crews (whose names were published by the Japanese newspaper Mainichi with the consent of their families) and ambulance rescue squads. A record of the victims' 83-day struggle with radiation sickness was published by the Japanese state broadcaster NHK with the agreement of the bereaved families. Crippled by radiation, the victims' bodies were unable to regenerate blood or organ cells and dehydrated as their skin peeled and shed, so that they died in terrible agony (NHK, A Slow Death: 83 Days of Radiation Sickness).

The fact that after the accident several tens of JCO staff members entered the storage tank with sandbags in order to arrest the criticality has remained unpublicized to date. According to a report from April 2000, the total number of victims of radiation was 667. From the start, however, the nuclear-fuel industry operatives, the nuclear-fuel company, and even the government and its officials, who were under pressure to act, showed an irritated distaste for this bothersome affair. Countless numbers of people have been killed by radioactive contamination and radiation exposure. Among them are the soldiers of the old Soviet army who penetrated the Chernobyl nuclear power facility and local children, who suffered high rates of thyroid cancer and leukemia; [23] the staff of the Hanford nuclear facility in the United States, among whom there was a sharp increase in radiation-related cancers; and communities surrounding the core processing facility at La Hague in France. If we also count nuclear explosions, the list of victims will include the crew of the Japanese fishing boat Dai Go Fukuryu-maru caught up in the Bikini Atoll hydrogen bomb tests and the victims of the Hiroshima and Nagasaki bombings, including those suffering their aftereffects down to the second and third generations.

There have been issues raised over the possibility that the fallout from the approximately 1,000 nuclear tests carried out in the desert of the US state of Nevada may have inflicted damage on the population of some 200,000 so-called down winders in Utah and Idaho, according to a 1964 survey report from the US National Institute of Health (NHK Hiroshima, March 15, 2014). Following the large-scale disposal of nuclear waste from the former Soviet Union and Britain in drum cans in the Atlantic Ocean, North Sea, Bering Sea, and Japan Sea, radioactivity leaked from the cans rusting in the seawater, and contamination spread to fishery resources. The direct dumping of nuclear waste into the Dover Straits by pipeline from France's La Hague nuclear processing facility caused contamination in bottom-dwelling cod and herring (Déchets: Le Cauchemar du Nucléaire, Bonne Pioche/Arte France, 2009), endangering the source of traditional dishes such as cod for fish and chips, Norwegian pickled herring, and salmon.

Politicians and academics who have interests in the nuclear power industry and no sense of history bear a heavy responsibility for having taken the lead in shaping the public's zone of concern and creating public indifference. In a small country like Japan, collusive relationships between politicians, government officials, and industry have bred a system of 'concealment and falsification', with the tendency to manipulate people's zone of concern similar to that of the Imperial General Headquarters during World War II. Although heavily defeated in the sea battles of Midway and Guadalcanal, it manipulated the population with information that claimed victory. The parents of soldiers who never returned were unable to hide their mistrust. The expansion of the zone of indifference relative to society, the individual, and the organization is marked compared to any other country. The unsafe aspects which people are apt to forget need to be imprinted in the social system, so that the structures and systems of the social organization are forced to compensate for them. This zone of indifference, according to C. I. Barnard, lies at the locus of conflict of "individual personality versus organizational personality" [24], and is outside the scope of the shared area which the citizen inhabits as an individual who is a working member of an organization and also a member of society and family.

Through the routinization of the individual's assimilation into society, risk and crisis are converted to issues of the social organization. Ultimately, however, disaster and accident victims are individuals. In terms of our powers of attention with regard to the variety of unsafe factors in our environment, the zone of concern of modern humans is contracting and the zone of indifference is constantly expanding. It seems that, in the process of flight from the lonely independence of citizen versus nation and individual versus collective who is located deep in the human psyche, there is a mechanism that amplifies unsafety (Fig. 7.6). The contraction of the individual's zone of concern increases the zone of indifference of the collective, and instead encourages the formation of a common zone of concern in the social organizations to which the individual belongs. This hierarchical inversion of the zones of concern and indifference between individual and collective is said to create the superordinate fiction, in other words power and authority. Within

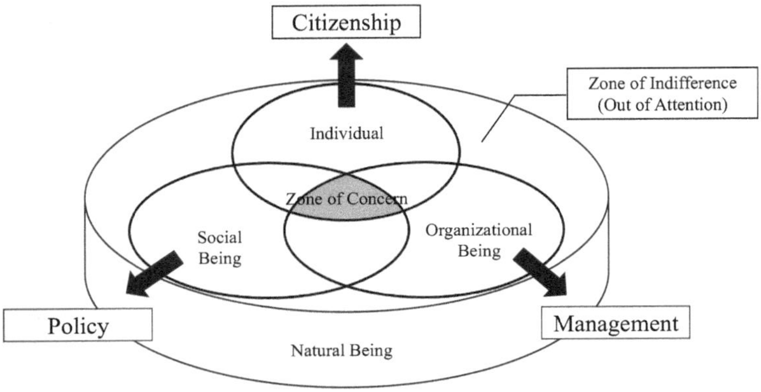

Fig. 7.6 Subconscious of human nature: organizational personality vs. individual personality. *Note:* Adapted from 'human hypotheses' from J. Piaget, E. Cassirer, A. Carrel, J. S. Bruner, W. Buckley, and N. Humphrey

H.A. Simon's concept of routinization, the focus of attention is constantly reviewed. This is based on the bounded rationality of every human being.

Assuming that an item were to be accepted into the zone of concern, one would need to trace back to find out who took charge of 'decision-making' over its management or the relevant policy, when, where, and with what right. On the other hand, when an item is accepted as part of the zone of indifference, the reverse applies. For instance, as in the Chisso Minamata disaster, there are countless examples where employees have tacitly accepted factory effluent that causes pollution (Chisso Minamata court case), government authorities have ignored environmental pollution to give priority to industry-promotion policy, the regulatory government bodies have prioritized commercial interests under the so-called convoy system of *Japan Inc.*, and shareholder interests have been emphasized at the same time as the very lives of staff, the local community, and other stakeholders have been trifled with.

When people take up a social or organizational role, they tend to take an attitude of "see no evil, hear no evil, speak no evil" with regard to unsafety factors. Freed from the weight of responsibility, they avoid friction with the collective. This is a negative avoidance mechanism which operates in every human from unsafety to risk and crisis [25]. However, the cumulative result is major disasters and accidents, the responsibility for which may rebound on the individual concerned. In the JR West derailment accident (Chap. 4), the former company CEO Mr. X, who was taken to court, had at one point been the general manager of the Railway Safety Division, and various responsibilities he had overlooked at that time were questioned [26]. Generally, in the case of a private-sector enterprise, business failure and organizational bankruptcy would ensue, and financial collapse is sometimes reported, even among government institutions in such situations. The result of careless management in both society and government is seen in the EU member-country Greece, in Japan and America with their massive fiscal deficits, and in the worldwide financial instability created by the Chinese government's manipulation of its currency. The potential for risk and crisis is such that we are walking on an unsafe minefield which may explode at any moment.

Even assuming that risk and crisis were to be cognized, in many cases the information is deliberately overlooked or ignored, for instance by employees or officers who prioritize the interests of the state or the organization, so that no preventive action is taken within one's own organizational culture [27]. When the individual has adopted a functional personality as an organizational member, the zone of concern is masked, and in many cases a pretense of not noticing will be maintained. As a result of assimilation into the social organization, the individual's powers of attention are dispersed, resulting in the expansion of the zone of indifference (Fig. 7.7). Instructions from government or the organization to place an item in the zone of indifference, as it has no particular bearing on the personality and beliefs of the recipient, will mostly be accepted without further comment. Conversely, in response to an instruction to activate the zone of concern, the recipient will be suspicious as to the basis for the instruction and will struggle between disbelief and trust.

At the time of the Fukushima Nuclear Disaster, residents did follow the government's recommendation to evacuate, but due to the handling of the processing of contaminated water after the accident, local residents and the Japanese public had increasing misgivings, becoming skeptical toward the power company and government and mistrustful of policy decisions and organizational decisions [28]. Regarding the handling of the contaminated water from Fukushima, it was reported that if confidence were not established and at the same time the locus of responsibility made clear, residents might refuse to obey recommendations issued by the government. It was also reported that the decontamination operations were inadequate. In this way, the zone of concern in connection with the Fukushima contaminated-water issue can be seen as a case where the fiction of state power and organizational authority, which had hitherto been accepted, began to peel away [29]. As with the collapse of the Berlin Wall, the reaction of local residents (especially farming and

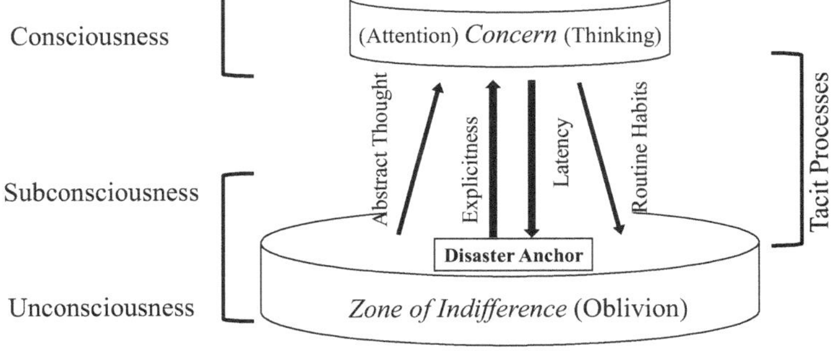

Fig. 7.7 'Disaster anchor' at the bottom of the zone of indifference. *Note:* Drawing based on construction from 'tacit dimension' by M. Polanyi and 'fiction of superior authority' by C. I. Barnard

fishing families) against the state, the government, and the power company over the contaminated-water issue was deep-rooted.

Local residents around the Fukushima nuclear power station waited for support from the national government and guarantee from the power company, but there was no improvement, and they were shocked into combative action by the inadequate response of politicians, government officials, and corporate executives and managers, who were anxious to save their own skins. Prime Minister Abe declared to the IOC Session at the time of the award of the Tokyo Olympics that "the Fukushima contaminated water issue is under control" [30]. Admittedly, given the industrious nature of the Japanese people, the decontamination operation may well prove effective, but the Japanese people are entitled to be a little uncertain as to whether the contaminated water was 'under control', affected as it is by natural phenomena such as tide, wind, and groundwater systems.

As shown in the figure below, we are essentially personality-based entities, but, as a result of participating in the social organization, we work from 9 to 5 o'clock for 5 days a week in return for a wage and other benefits. In this context, we are required to fulfill a functional role in line with the organizational purpose, and our thinking and behavior are restricted. In the personification process of transition from the individual personality to the organizational personality, we 'depersonalize' ourselves, and become a part functioning as a constituent member of the organization. On the other hand, although both society and the organization might be thought to originally be non-personality based, the 'quasi-personality' delegated to it by its members permeates the whole. The sum of the transferred aspects of people's thinking and behavior constitutes the whole personality. The personified society and organization commissions to the individual responsibilities in line with the needs of the collective, but it is in few cases that the authority commissioned is commensurate with the responsibilities delegated to the individual [31]. In other words, authority and responsibilities are in social imbalance (Fig. 7.8).

Many modern humans, by emphasizing their belongingness and assimilation to the social organization, may obtain a deceptive outward appearance of safety and peace of mind. In recent years, we have been increasingly assailed by a series of individual disasters and accidents. The origin of unsafety is in the deceptions and delusions which arise from the routinization of our evasion of confrontation with the collectives of state and society. We do not have as much free will as we imagine, as the freedom of our behavior and will is limited by the scenarios of our everyday activity and affairs. In J. Diamond's example of the 'two horses' cited below, which of them to choose is a matter to be decided not by politicians and government officials, nor by corporate executives, but by the individuals based on human nature who have left behind their functional personality, and us the citizens who are concerned for our descendants and the environment [32]. Whether capitalism or socialism, if we lose sustainability, the cumulative result will be our rejection by the global environment.

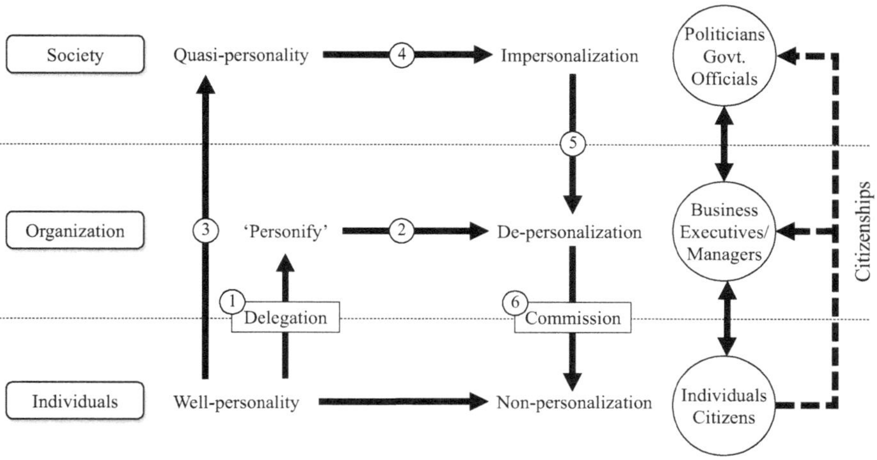

Fig. 7.8 Invisible informatics of personification in social organizations. *Note:* 'Personify' as the cognitive depths psyche of individual mental processes participated in the social organizations

7.3 Personifying the Environment: Deterrent of 'Unsought Consequences'

Who will speak up for the environment? Should it be the voice of developed countries or developing countries? Or perhaps politicians and business people? Or members of the public and scientists? This uncertainty underlines these questions, when speaking up as it were on behalf of the environment, which cannot speak for itself. Although the general theory is shared, there is disagreement on the details, leading to extreme diversity of opinion. According to L.R. Brown, mass-consumption society imposes burdens on the global environment and ends up working against human sustainability [33].

We cognize and learn about the environment through the environmental changes that we apprehend, for instance, abnormal weather patterns, floods, and other damage caused by natural disasters and the threatened extinction of plants and animals due to changes in the ecosystem. Speaking in various situations on behalf of the environment, which cannot have its own personality, we actually end up personifying it. Treating the environment as if it were a concept embodied by the ecosystem, people interpret it differently according to their experiential realities and cultural and linguistic particularities. There is of course a difference of degree, but unless we refrain from the personification of the global environment, unless someone speaks up for the environment, and unless we take seriously global warming, CO_2, water resources, environmental hormones, chemical contamination, food crisis, crop failure, and population explosion, it will be our own extinction which forces us to notice. It is when natural disasters and unnatural disasters, which we are convinced are all someone else's problem with no impact on ourselves, threaten us imminently that we learn of the danger lurking in the environment.

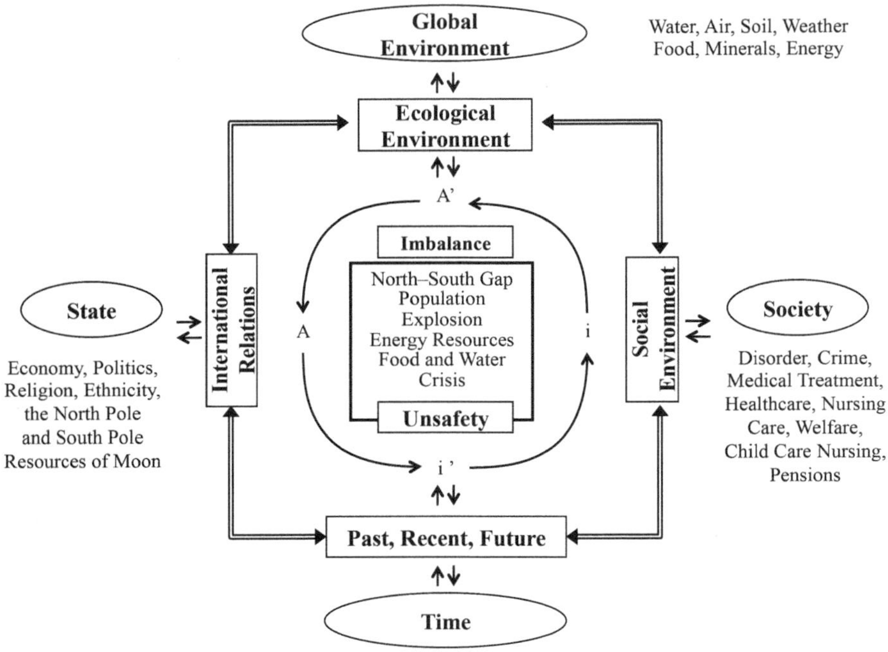

Fig. 7.9 Five-layer model of unsafety (risk, crisis, accident, disaster, and catastrophe). *Note:* Drawing based on consideration from Saving the Planet and State of the World 2000 by L. Brown. 'Layer model' concept based on a suggestion by M. von Zedtwitz at IMD International, Lausanne 2002

The environment is subject to the limitations of the world's many and various contexts and is a conceptual construct mediated through elements of the physical, cosmic, geographical, chemical, biological, ecological, social, economic, political, ethnic, cultural, philosophical, and so on.

Meanwhile, as an example of unsafety, it is said that American society is so unbalanced that 1 % of the population owns 99 % of the wealth (NHK, BBC, DR, ITVS, SVT, ZDF/Arte VPRO Steps International, 2012). Due to the recession, more and more of the middle class are becoming destitute. The United States, the world's leading economy—where the number of guns in ownership is greater than the nation's population and there are frequent random shooting incidents involving schoolchildren—is an area of 'unparalleled danger' in historical, geographical, and social terms. Unsafety, which is not limited to the United States but is spreading worldwide, is the subject of the five-layer model in Fig. 7.9.

A. Lovins says that, to realize his 'sustainable society', we need to wean ourselves off conventional fossil fuels [34]. That will require a switch from the previous high-density large-scale urban system to build a smaller-scale decentralized social model. Even now, the fossil-fuel-dependent society is reaching its limits, with the rapid increase in CO_2, methane, and other greenhouse gases sparking abnormal weather patterns. Lovins proposes a self-sufficient small- to medium-scale decentralized model adapted to the size of the community. Seen historically, many recent wars in developed countries have been caused to no small extent by the struggle over fossil fuels.

Under the sustainability project in the city authority of Portland in the US state of Oregon, the deterioration of the water and atmospheric environments due to past industrialization was identified as an issue and city resources were focused on initiatives to create a 'clean green city' [35]. The city authorities encouraged switching from car to bicycle and promoted bicycle lanes and the use of trams, resulting in a halving of exhaust-gas emissions. Also, switching to a policy of food self-sufficiency, the authorities limited inward imports from distant locations, and restricted car use within the city, which succeeded in cutting CO_2 emissions by 23 % (Passport to Knowledge Productions, 2012). If Portland city's initiatives could be rolled out to the whole of the United States, it is estimated that the world's oil consumption could be reduced by 25 % due to less road transportation, with benefits from reduced transportation costs, and reduced environmental costs in terms of CO_2 emissions. Even if energy sources were to become depleted and fossil fuels and imported foods unavailable, an autonomous society would be able to sustain urban functions and communities as wasteful transportation costs and CO_2 and other forms of energy consumption would be reduced. Moreover, Portland citizens and city authorities work to regulate the operations of commercial enterprises under a sustainable approach that emphasizes the life of citizens rather than economic growth as under traditional capitalism. In the 40 years since 1974, citizens have thus worked toward an 'eco-civilization'.

In the German village of Feldheim, a municipally operated wind-powered electricity-generation facility, which has attracted interest from the world's energy scientists, represents a new initiative in renewable energy [36]. Under this electricity-generating project, led by the local community, the electricity-transmission grid of the electric power company was bought for around US $600,000 and an experiment in energy self-sufficiency launched with the aim of supplying the consumption needs of the local community. The investment required setting up the generation facility, and the costs of the transmission grid were shared among residents (around $4,000 per household), who have benefited from a saving of around 50 % on yearly household electricity costs compared to the previous supply by the electric power company. In regular electricity grids, attenuation during transmission results in a 50 % wastage, but locally self-sufficient systems of electricity generation and transmission are effective in reducing this wastage. When there is no wind, the facility can switch to backup thermal power generation, using not fossil fuels but biogas electricity generated with methane gas from livestock. With biogas electricity generation, the fermentation temperature is 80 °C, allowing use as a source of hot water, so that yearly heating and lighting costs have shrunk from around $2,700 to around $1,600. Achieving both sustainability and renewability, this can be seen as a successful example of small-scale decentralized provision. The features common to Portland in America and Feldheim in Germany are that, whether for electricity generation or for transmission and distribution of goods, the local community took the lead rather than relying on a

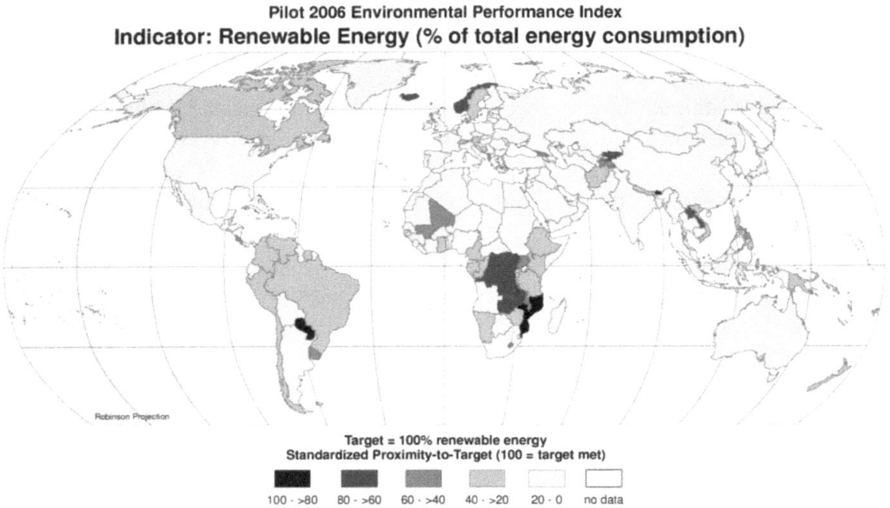

Fig. 7.10 Worldwide renewable energy resources. *Source:* NASA, Socio-economic Data and Applications Center (SEDAC), by NASA permitted Website Policy (*Publicity*). *See also,* Yergin, D., The Quest. [37]

'market economy' based on private-sector enterprises, and that the core focus was on sustainability rather than capitalism. This is a visionary approach. Figure 7.10 presents a world map of the potential for renewable energy resources as researched by NASA and D. Yergin; [37] and Table 7.2 information from a WWF report and forecast on renewable energy by source.

Now that fossil fuels are outdated and the time to shift to renewable energy has come, it has been said that branching out into new energy sources is an opportunity for evolution toward the society of the future. One place that has attracted attention for its shift to a new energy source that utilizes geographical features is Iceland. With a state target of 2050, the aim is to create a non-fossil-fuel-based society through geothermal electricity generation (NHK, November 2, 2012) [38]. Already, in addition to electric power supply from the country's four geothermal electricity-generation stations, hot water from geothermal sources is supplied to all households and widespread use is made of steam heat for cooking and saunas. In the future, to encourage widespread use of hydrogen-powered vehicles, the number of hydrogen refueling stations is to be increased. In the United Arab Emirates, meanwhile, solar electricity generation in the scorching desert has been used since 2013 to generate electricity from hot water using reflective plates. Abu Dhabi has a 10-year plan for a small-scale, decentralized, clean-energy model city in the desert to be known as 'Masdar City' (Saint Thomas Productions, 2012) [39]. For initiatives of this kind in various countries to achieve sustainability, a precondition is the approach of an 'autonomous' society. This assumes a recycling society where, metaphorically speaking, each person only brings as much luggage as he or she can carry, and renounces the greedy wish to have more. The aim is to reuse resources and remove

Table 7.2 Global energy provided by source and year

Source	2000	2010	2020	2030	2040	2050
Total electricity (EJ/a)	45.7	60.0	71.9	85.7	22.0	127.4
Wind power: on-shore	0.2	1.4	6.7	14.3	3.4	25.3
Wind power: off-shore	0.0	0.0	0.5	1.3	0.3	6.7
Wave and tidal	0.0	0.0	0.0	0.1	16.9	0.9
Photovoltaic solar	0.0	0.1	0.7	6.5	13.7	37.0
Concentrated solar power	0.0	0.1	0.6	3.9	14.8	21.6
Hydropower	7.9	11.3	13.4	14.4	3.4	14.9
Geothermal	0.1	0.3	0.7	1.7	1.7	4.9
Biomass	0.0	0.0	0.0	0.0	5.4	16.2
Coal	18.2	21.5	14.8	10.0	20.1	0.0
Gas	8.6	14.0	25.6	28.3	0.5	0.0
Oil	4.2	3.1	2.5	1.4	0.5	0.0
Nuclear	6.5	8.2	6.5	3.8	1.2	0.0
Industry fuels and heat (EJ/a)	63.7	79.1	82.3	74.6	63.0	59.0
Concentrated solar: heat	0.0	0.0	0.1	0.4	2.6	8.8
Geothermal	0.0	0.1	0.2	0.6	1.6	2.9
Biomass	1.0	6.1	16.9	31.3	40.7	34.8
Fossil fuels	62.7	72.9	65.0	42.2	18.0	12.5
Building fuels and heat (EJ/a)	77.7	86.0	87.4	67.8	47.4	24.1
Solar thermal	0.0	0.7	3.3	11.9	16.0	12.6
Geothermal	0.2	0.5	1.5	4.1	10.5	8.4
Biomass	33.4	33.2	29.2	14.2	10.2	3.1
Fossil fuels	44.1	51.6	53.5	37.6	10.6	0.0
Transport fuels (EJ/a)	86.2	102.6	111.6	91.3	62.3	50.8
Biomass	0.7	4.8	12.9	29.7	45.7	50.8
Fossil fuels	85.5	97.8	98.8	61.7	16.6	0.0
Grand total (EJ/a)	273.4	327.6	353.3	319.4	276.2	261.4

Source: The Energy Report—100 % Renewable Energy 2050, WWF, 2011

the wasteful aspects of prosperity: in other words to shift to a society that knows when it has enough.

As a pointer toward a sustainable society, J. Diamond says that, historically, not all civilizations collapse, and cites the example of Japan under the Tokugawa Shogunate, where forest resources were continuously controlled for around 300 years [40]. In the Edo period, forest resources and rice were the main products of the Japanese economy and the role of the forest was important, including supplying firewood for cooking and timber for wooden buildings and castle construction. The Edo Shogunate regulated timber felling, thus maintaining sustainable forest resources. The state system of the Edo Period was actually a class system, 'shi–nō–kō–shō', which divided the population in a hierarchy from warriors (shi: samurai as knights), through nō (farmers) and kō (craftsmen), down to shō (merchants). To be more exact, above the warrior class was a court society of royalty and aristocracy around the Emperor. Modern Japan, on the other hand, leaving the Imperial Family and aristocracy aside, has changed into a society with a hierarchy from merchants, through craftsmen and farmers, down to warriors, 'shō–kō–nō', where the merchants—corporations and the rest of the business world—exploit the skills of the craftsmen and use the resulting profit to fund the warriors: business-friendly Diet congress politicians and government officials in charge of administration in the public and private sectors. In this way, business management and government policy are frequently in a collusive relationship. The self-sufficiency rate achieved by the farmers is a mere 30 %, so the country is dependent on overseas trade for energy and food resources and is a long way from being the kind of autonomous society that aims for self-sufficiency through its own produce. In an age when a multipolar and decentralized arrangement is said to be the most secure, the increasing unipolar concentration around Tokyo appears to run counter to the worldwide trend toward a sustainable society.

As outlined above, the city of Portland has been working since 1974 on a sustainable-society plan. As part of this, it has drawn limit lines—allowing amoeboid movement with changing conditions—for urban growth throughout the city area as a 'mechanism to limit materialism'. Within the limit line of Portland only, the city authorities delineated commercial and residential areas, drafted regeneration plans for areas outside the limit lines, and adopted a policy of not permitting development activity. The area outside the limit line is earmarked as a rezoning green division exclusively for forest and farmland. The decision-making mechanism of the city administration includes a 'five-person committee system' [41] set up to seek the opinions of citizen stakeholders and representatives of a range of other groups, as well as a policy–decision system which takes into account the opinions of neighborhood associations in each town. The actual policy decisions are taken under a direct-democracy model in which effective decision-making lies not with the mayor or city council but with the five-person committee of specialists who take immediate decisions in response to citizen proposals, thus realizing an effective sustainable-society model (Fig. 7.11).

Professor Steve Johnson (Portland State University) focused on the fact that the five-person committee took statements from 350 groups and 4,000 associations in Portland city, meaning that citizens participate directly in politics, and commented that "eco-civilization is being propelled by sustainability-focused reforms in the social system driven by citizen power, in which policy decision equals citizen decision" (Passport to Knowledge Productions, 2012). This kind of 'cooperative citizenship' made an appearance in Japan at the time of the Great East Japan Earthquake, when local communities provided mutual assistance ahead of the 'too-little too-late response' of the administrative authorities and central and local government, showing a glimpse of an autonomous society where citizens engage in self-motivated and continuous cooperation. The Portland city authority also attaches importance to stakeholders with regard to the local economy and environmental protection, operates a policy of tax incentives for businesses installing rooftop waterproof solar panels, takes measures to boost the local economy through

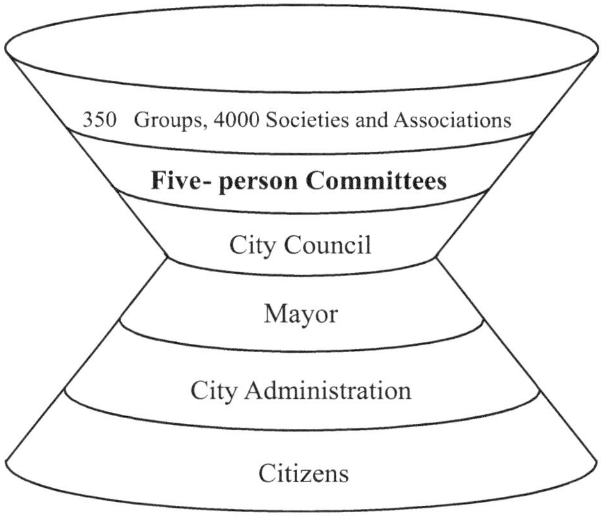

Fig. 7.11 Case study: Portland city stakeholders in USA. *Note:* 'Five-person Committees' as a decision-making nucleus for the welfare of Portland

'green enterprises' that sell eco-products and create employment, and supports the interests of local eco-businesses and the simultaneous development of the local economy.

According to A. Lovins, the "market economy is a good tool and a good assistant, but if not used smartly, it ends up creating a market society", in which it becomes difficult to reflect citizen power directly in policy decisions. M. J. Sandel foresees the possibility that social conventions will be lost and 'moral dilemmas' reached under the dominant culture of money worship centered on material acquisition which has been encouraged by modern capitalism as the market economy has developed into the market society in the period since the Industrial Revolution [42].

7.4 Disaster Management and Crisis Sciences

J. Diamond tells a fable of two horses: will we choose the 'horse of destruction' representing environmental ruin through excessive consumption, or the 'horse of sustainability' representing a sustainable society where material greed is curtailed? He says we are at present poised between the two [43]. D. Yergin asserts that the development of solar and wind power will take 40 years, but that the renewable energy of biofuel (biomass) is usable as aviation fuel and that it will be possible in the future to produce oil and electricity for transport through biological photosynthesis, realizing the sustainability of an autonomous society.

On the other hand, there are many commentators who point to the limits of modern capitalism with its collusive relationships between business management seeking corporate profit and government policy. Contrasting with occupation and private possession, the principles of common use are important for the global environment. Although we may draw borders on the surface of the globe, the groundwater systems that flow beneath us and the air above us cannot be divided up. There is essentially no such thing as complete occupation. Private occupation of land does not extend to freedom to exploit the mineral resources and water resources beneath it, and possession of the air is impossible. The extension of the modern capitalist concept of ownership signifies pollution and destruction in the name of a kind of environmental manufacturing based on an accelerated 'short-term consumerism' which leaves us unable to guarantee the environment for the next and successive generations. As a result, environmental assets under private ownership, whether those of an individual or of a business enterprise or other organization, are all exhausted within one generation. This perversion toward individual wealth accumulation is a double-edged sword of capitalism, increasing the threat to the sustainability of society.

Under modern capitalism, the underground mineral resources of land under ownership are exhausted in one generation. When we look at the historical reality that factory production has polluted the groundwater systems below and the air above, it appears that private possession amplifies human material greed, which then knows no bounds and accelerates the delusion of wealth accumulation. In the final analysis, the actual reality of private possession of assets is not occupation but exclusive holding or exclusive use. It is unreasonable to say that ownership confers the right to threaten life and to complete freedom of use of surface rights and the assets owned. If we assume temporal continuity, with sustainability of our descendants and the ecosystem, then we are merely exercising temporary common use. To pass the global environment on to the next generation and promote sustainability, common use should take precedence over ownership and exclusive holding. To leave room for life, there is an increasing need to rethink the destructive factors arising from the excesses of capitalism and to aim for a revolutionary change within global eco-revolution to a sustainable social structure based on common use. Curtailment of expanded material greed and social innovation founded on sustainability rather than capitalism are perhaps self-evident necessities.

The negative legacy of human inhabitation consists not only of CO_2 and environmental hormones such as polychlorinated biphenyls (PCB) and dioxins. Orbiting the earth at a speed of 7–14 km a second, there are apparently 100 million items of what is called space debris, which pose an increasing danger to space development and exploration according to the 'Kessler syndrome' [44]. There are already countries disputing rights to the dominion of the moon. This extension of the state's ownership rights and exclusive-use rights into space is something at which we can only despair. The activity cycles of galactic space and the solar system have an impact on global climate change. Within the global environment, which is currently in an interglacial period between ice ages, it can at least not be ruled out that they are connected with ocean warming from human emissions of CO_2 and thermal effluents from nuclear reactor cooling and the resulting Big Melt: that is, the melting and release of methane hydrate from permanently frozen soil and the seabed, releasing methane gas with at least 20 times the greenhouse-gas effect of carbon dioxide. According to NASA, galactic supernova explosions

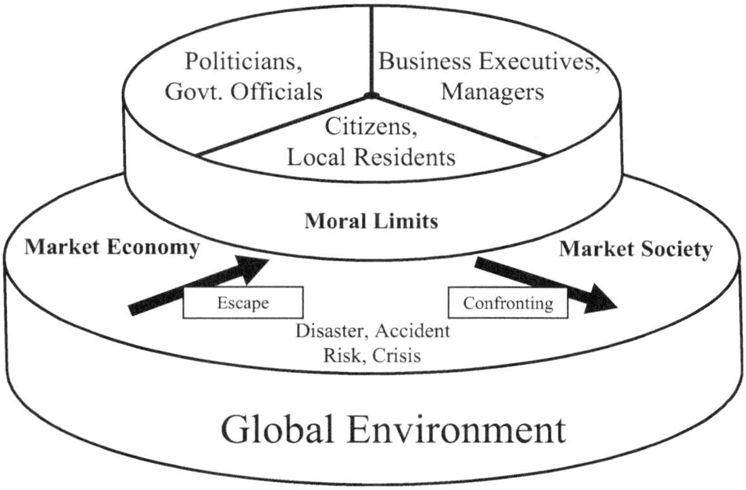

Fig. 7.12 Disaster management and crisis sciences by confronting unsafety. *Note:* Drawing based on *Escape from Freedom* by E. Fromm and *Justice: What's the Right Thing to do* by M. Sandel

increase the quantity of cosmic rays and the amount of cloud and create the conditions for an ice age. Going forward, regardless of whether the result is global warming or cooling, abnormal weather patterns will occur frequently, and crop failure is feared. Coinciding with the global population explosion, this brings increasing risk of a food crisis. The resolution by *crisis sciences* of the total issue of unsafety [45], which now encompasses natural disaster and human-made disaster, can be seen as a test that humankind must now face (Fig. 7.12).

The belief that the global ecosphere (Gaia) is limitless is collapsing, and the advantages that bring economic and social safety and stability are being lost. Modern reality with its polarization between the environment and humans is causing the zone of indifference to expand and driving people away from their free will as natural beings. This eventually becomes the wellspring of environmental destruction. The appearance of assimilation can be seen as a weakness whereby one seeks to attain mental tranquility through confrontation avoidance by delegating to others not only one's individual right to self-determination but also one's responsibilities. In other words, the 'great human paradox' is precisely that we are inclined to prefer over safety an immediate convenience which is unsafe. As a result, in the face of the danger of environmental destruction latent within us, we seek release by choosing to escape from the issue into the zone of indifference. For instance, modern humans, out of fear of the environmental hormones accumulating in our organs [46], find mental tranquility in a pseudo escape (*see also*, Fig. 7.5: cognitive zone of indifference). In response to the threat of toxic substances building up in our bodies, we achieve peace of mind in return for the deception of cloaking ourselves in indifference. As Silver indicates, bio-enterprises which genetically manipulate the organs and DNA of the human body and cloning, who are destined to suffer illness and death, are nowadays causing ethical unsafety [47]. Sorrow and transience: these are the hallmarks of humans under the sway of our times.

By analogy with E. Fromm's *Escape from Freedom* [48] and M. Sandel [49], the present chapter analyzed the human condition whereby, despite cognition of various signs of disaster, we take no action. There are many modern humans who have gaps in their conscience, for instance when they tacitly accept illegal factory effluent or remain silent when confidential documents are passed on or bullying takes place in schools. Whether it is F. Tönnies's 'Gesellschaft or Gemeinschaft' [50], 'human weakness' and the attribute of conforming with the collective, whereby the collective is transcribed onto the individual, arise from the inability to live without belonging to the social organization. What can be concluded from this is that, in the process of encountering risk and crisis and the unsafety which brings no benefit, evasive behavior in the style of "see no evil, hear no evil, speak no evil" and mental evasion operate, lowering the immune strength of the social organization. Individuals with an allocated societal role, by masking their professional responsibilities, render unsafety invisible. Here, the limited nature of technologies and systems based on profit and of capitalism focused on wealth and material goods is exposed through breakdown and 'moral bankruptcy'. The situation surely demands a sustainable approach that does not reduce people's degree of satisfaction and happiness but preserves it for the next generation. The reality is that the accumulation of human-made disasters studied in this book poses the danger of precipitating a natural disaster of tragic proportions and ending in catastrophe. Rather than focusing on our responsibilities as members of organizations, it is to be hoped that we will adopt a perspective founded on our responsibility as global citizens living in Gaia and on our responsibility to the next generation.

References

1. Diamond, J.M., *Collapse: How Societies Choose to Fail or Succeed*, Viking Penguin, 2005.
2. McNeill, W.H., *A World History*, Oxford University Press, 1998.
3. Fromm, E., *Man for Himself: Enquiry into the Psychology of Ethics*, Rinehart and Company, 1947. *See for example*, Torahiko Terada.
4. Thom, R., Zeeman, E.C., Ushiki, S., Sawa, T., *Keitai to kouzō—katasutrofu no riron* [*Catastrophe Theory*], Misuzu Shobō, 1977.
5. Leiserowitz, A., "Climate change risk perception and policy preferences: The role of effect, imagery, and values", *Climate Change* 77: 45–72. 2006.
6. Cribb, J., *The Coming Famine: The Global Food Crisis and What We Can Do to Avoid It,* University of California Press, 2010.
7. Lloyd, C., *What on Earth Happened?: The Complete Story of the Planet, Life and People from the Big Bang to the Present Day*, Bloomsbury Publishing, 2012.
8. Heinrich, H.W., *Industrial Accident Prevention – A Scientific Approach*, 1931.
9. Reason, J., "The dimensions of safety" in Patrick, J. (ed.) *Cognitive Science Approaches to Cognitive Control*, Third European Conference, Cardiff: University of Wales, 1991.

10. Pierce, J.R., *Symbols, Signals, and Noise: The Nature and Process of Communication*, Harper & Bro., 1961.
11. Norman, D.A., *Learning and Memory*, Freeman and Company, 1982.
12. Polanyi, M., *Personal Knowledge*, Routledge & Kegan Paul, 1958.
13. Simon, H.A., *Administrative Behavior: A Study of Decision–Making Processes in Administrative Organizations*, Macmillan Publishing, 1945, pp.210–212.
14. Barnard, C.I., *The Functions of the Executive*, Harvard University Press, 1938, pp.167–170.
15. Simon, H.A., *op. cit.*, pp.80–84.
16. Ashby, W.R. *An Introduction to Cybernetics*, London: Chapman & Hall, 1956.
17. Polanyi, M., *Tacit Dimension*, Routledge & Kegan Paul, 1966.
18. Ayer, A.J., *The Foundations of Empirical Knowledge*, Macmillan, 1940.
19. Barnard, C.I., *op. cit.*, pp.163–165.
20. Weick, K.E., "The vulnerable system: an analysis of the Tenerife air disaster" in Frost P.J. et al., (eds), *Reframing Organizational Culture*, London: Sage Publications, 1991.
21. Atsuji S., et al., "Organizational Disaster by Implicit Systems Errors: A Case Study of the JR West Accident", *IFSAM*, 2012.
22. Barnard, C.I., *op. cit.*, p.270.
23. Yablokov, A.V., Nesterenko, V.B., Nesterenko, A.V. and Preobrazhenskaya N.E., *Chernobyl: Consequences of the Catastrophe for People and the Environment*, the New York Academy of Sciences, 2009.
24. Barnard, C.I., *op. cit.*, pp.88–89.
25. Glendon, A.I. and McKenna, E.F. *Human Safety and Risk Management*, London: Chapman & Hall, 1995.
26. *See for example*, Atsuji et al., *JR Accident 4.25*, Kansai University, 2007.
27. Weick, K.E., "Organizational culture as a source of high reliability", *California Management Review*, 29, 1987, pp. 112–27.
28. Atsuji et al., "The Fukushima Catastrophe seen as a Malfunction of Organizational Systems: Nuclear Fade–out and Hazard Maps for the World", *General Systems Bulletin*, Vol. XXXXI, 2012.
29. Atsuji S. and Cook, N.D., "Fukushima Nuclear Catastrophe 3.11: System Pathology of Social Organizations", *ISSS*, 2011. *See also*, Atsuji et al., "Sustainable Decision–Making Following the Fukushima Nuclear Catastrophe", *IFSAM*, 2012.
30. *See*, The ASAHI SHIMBUN, Doubt cast on Abe's assurance to IOC about Fukushima leaks, 10th Sep. 2013. *Available* at: https://ajw.asahi.com/article/0311disaster/fukushima/AJ201309100071. *Accessed* March, 2014.
31. Barnard, C.I., *op. cit.*, pp.170–172.
32. Diamond, J.M., *Guns, Germs, And Steel: The Fates of Human Societies*, W. W. Norton, 1997.
33. Brown, L.R. et al., *Saving the Planet: How to Shape an Environmentally Sustainable Global Economy*, Worldwatch Institute, 1991.
34. Lovins, A.B. et al., *Saiseikanō Energy ga Hiraku Mirai [A Future Built on Renewable Energy]*, Iwanami, 2013.
35. *See*, Portland. *Available* at: http://www.portlandoregon.gov/. *Accessed* May, 2014.
36. *See*, Feldheim. *Available* at: http://www.neue-energien-forum-feldheim.de/. *Accessed* April, 2014.
37. Yergin, D., *The Quest*, Penguin, 2011.
38. *See*, National Energy Authority of Iceland. *Available* at: http://www.nea.is/. *Accessed* May, 2014.
39. *See*, Masdar City. *Available* at: http://masdarcity.ae/en/. *Accessed* May, 2014.
40. Diamond, J.M., *Collapse: How Societies Choose to Fail or Succeed.*
41. *See*, Portland. *Available* at: http://www.portlandoregon.gov/. *Accessed* May, 2014.
42. Sandel, M.J., *Let's Talk about Justice in the World and a Special Lecture in Tohoku University*, Hayakawa Publishing, 2013.

43. Diamond, J.M., *The World Until Yesterday: What Can We Learn from Traditional Societies?*, Viking Penguin, 2012.
44. Kessler, D. J. and Cour–Palais, B.G., "Collision Frequency of Artificial Satellites: The Creation of a Debris Belt", *Journal of Geophysical Research* 83, 1978.
45. Kineman, J.J., "Crisis sciences for sustainability", *Program & Abstracts 2015,* (ISSS Journals), pp.166–167.
46. Colborn, T., Dumanoski, D. and Myers, J. P., *Our Stolen Future: Are We Threatening Our Fertility, Intelligence, and Survival?*, Spieler, 1996.
47. Silver, L.M., *Remaking Eden: How Genetic Engineering and Cloning will Transform the American Family*, Avon Books, 1997.
48. Fromm, E., *Escape from Freedom*, Rinehart and Company, 1941.
49. Sandel, M.J., *Justice: What's the Right Thing to do*, Farrar, Straus and Giroux, 2009. *See also*, Sandel, M.J., *What Money Can't Buy: The Moral Limits of Markets*, Farrar, Straus and Giroux, 2012.
50. Tönnies, F., *Gemeinschaft und Gesellschaft*, Leipzig: Fues's Verlag, 1887.

Chapter 8
Crisis Sciences for Sustainability beyond the Limits of Management and Policy

In his work *Collapse*: *How Societies Choose to Fail or Succeed*, J. Diamond analyzes the collapse of the Easter Island and Maya civilizations, establishing that population increase resulted in food shortage and forest depletion and that these environmental changes brought about the destruction of cities. In another of his works, *Guns*, *Germs*, *and Steel*, he traces the development of civilization and technology from the steel blade through the gun to nuclear power and points to human-made disaster with population explosion, differential gaps and wars as a cause of civilizational collapse [1]. Among the triggers of collapse cited are biological factors such as plague and HIV infection and infectious pathogens spread from livestock food sources (smallpox, measles, tuberculosis, influenza, whooping cough, and malaria) [2].

The universe has a history of 13.8 billion years and the earth of 4.7 billion years, during which the globe has alternated between 'snowball' periods, when it is completely frozen, and 'big melt' periods. Currently, it is in an interglacial phase. Having outlived the Ice Age and spread across the globe, humans survived by progressing from a hunter-gatherer to an agrarian society. As a result, humankind established a sustainable civilization and culture that lasted 200,000 years and laid down a rich store of survival wisdom [3]. Today however, human technology and civilization are changing the natural ecosphere, and the greed of humankind as a whole, heading toward a population of 10 billion in 2050 by United Nations Report, has begun to exceed the limits permissible to maintain the balance of nature.

This chapter, carrying on from the causal relationship between natural and human-made disaster argued in the previous chapters, examines the disaster synchronicity that leads to catastrophe, and points to the risk of accidents and disaster arising from collusion between the social functions of business management and public policy. It goes on to review 'citizen power' as a social function to prevent the universal human tragedy of disasters and accidents, and as a movement from 'capitalism toward sustainism', discusses 'eco-civilization' [4] as a form of symbiosis with the environment.

S. Atsuji, *Unsafety*, Translational Systems Sciences 7,
DOI 10.1007/978-4-431-55924-5_8

8.1 Bounded Rationality: Limits of Management and Policy

Of the world's 194 countries, 105 nations located across the globe have seen war, conflict, coup d'état, riots, terrorism, or other upheavals since World War II [5]. History teaches us that the roots of conflict lie in competition for food, water, minerals, energy, and other resources; the two world wars were fought mainly over the energy resources of petroleum and coal [6]. The origins of human-made disaster are various and lie not only in the securing of resources but also in economic, political, social, and cultural clashes such as religious and ethnic conflict, and territorial struggles to secure arable land for food production. As shown in the map below, the fear and stress, anxiety and isolation, and hardship and discrimination that war and conflict bring are the same under any social system, whether capitalism or socialism. In addition to the imbalance these bring, the current population explosion toward a figure of 7.2–10 billion (estimated by the UN) is accompanied by intensifying conflict over water resources, arable land, fishing grounds, and mineral and energy resources [7]. Given the added effect of CO_2, small particulate matter, and other forms of atmospheric pollution, current levels of unsafety relating to the water, air and soil are increasing exponentially, not helped by the fact that rights are being asserted not only to airspace but also to the polar regions and even the moon (Fig. 8.1).

The world is growing unsafe not only because of natural disasters in the form of extremes of heat and cold, frost damage, drought, and flood. The crop failure

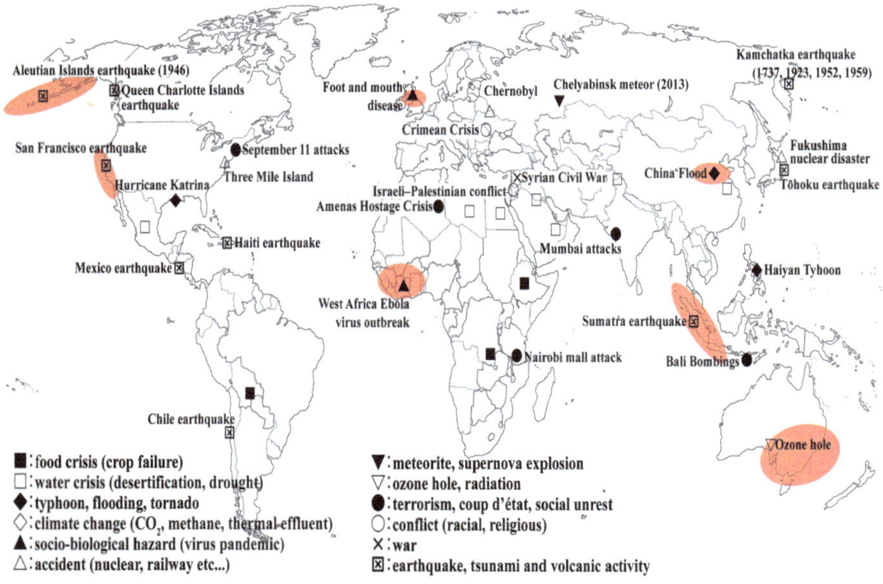

Fig. 8.1 World hazard maps 2015. *Note:* Mapping based on a discussion of risk/crisis at IFSAM Paris 2010 using data from United Nations (*Publicity*)

associated with these natural disasters, combined with population increase, leads to food crisis, and threatens the very survival of humankind [8]. On top of the natural disasters, there are also many human-made disasters brought on by war, conflict, and other causes of biological, economic, social, and political crisis. Another category of unsafety is risk from human-made factors such as expanding contamination with disease due to rapid population increase, waves of regional tension due to war and civil unrest, and racial discrimination arising from the growing wealth gap [9]. The origins of human-made disaster in modern society may not be susceptible to control by reason and intellect, with emotion and greed the eternal font of unsafety. To give a definition of unsafety, it can be generalized as social, economic, political, psychological, cultural, ethnic, geographical, or religious imbalance. Humankind may be no more than an essentially unsafe water- and carbon-based organic system.

The idea of a safety that everyone in the world can aspire to together is an illusion, associated with the contradictory dilemma that one person's safety may be another person's unsafety. Today's society has pursued material wealth, but people are becoming skeptical of a lifestyle where the rich become increasingly richer, while poverty is stubbornly self-perpetuating [10]. Such is the Japan of today, where the failure to achieve food self-sufficiency or provide a social insurance system is compounded by high living costs and tax levels and the additional burden of radioactive contamination and atmospheric pollution, causing anxiety and a gradual squeeze on public living standards. At the same time, the corporate disasters and organizational accidents hidden in the shadow of economic growth offer examples in which the business world alone has been protected and the victims abandoned, provoking the beginnings of doubt as to whether public safety and peace of mind can be guaranteed [11]. The effect of this gradually infiltrating unsafety is sensed by the next generation, leading many of the young to lose hope. Against this background, adopting the scientific study of unsafety is a perfectly natural means of preventing tragedy in the near future.

The military spending of the emerging nations is on an upward spiral. As shown in the map below, there is annual growth in the proportion of the GNP of the BRIC countries accounted for by the arms trade, which is already worth more than the grain trade, causing international tensions among regional neighbors and threatening safety. In February 2014, the winter Olympics were held at Sochi on the Black Sea coast, while next door in the Ukraine, the government was in collapse and Russian involvement in the Crimea Peninsula was shaking up the international community. Somewhere in the world is always in a state of war, conflict, coup d'état, riot, or terrorism: [12] as shown in the map, 105 of the world's 194 countries have seen outbreaks of war or conflict since World War II (The World Bank, 2011) [13]. The north–south wealth gap is a major factor in the continuously increasing unsafety arising from human-made disaster, but others include religious conflict (between Christianity and Islam), ethnic conflict, and the competition for energy resources around the Senkaku Islands in the East China Sea and the offshore oilfields in the South China Sea.

As shown in Fig. 8.2, although a country may have a nominally democratic form of government and be under the rule of law, there are many historical examples

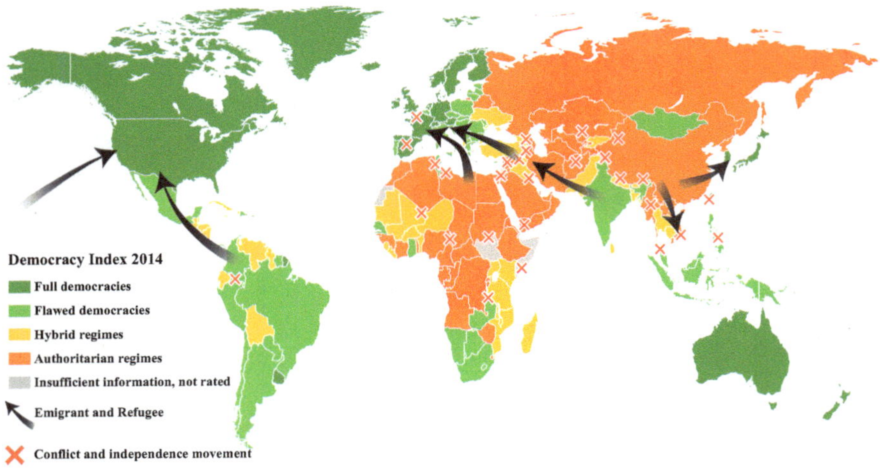

Fig. 8.2 Global unsafety: EU refugee crisis, democracy gaps, conflicts, and wars 2015. *Note:* Mapping by R. Fujimoto and S. Atsuji based on *Map of Freedom 2014* (Freedom House)

where the governing structure is characterized by a supra-legal element which determines the will of the nation. This may consist, for instance of (a) religious dogma, (b) plutocratic or status-based influence, or (c) rule by a particular ethnic or family group, or influences on the democracy of economics, religious, party-political, racial, military, environmental factors and refugee flows.

The reality of illegal decision-making and action at a premodern social level is not restricted to the Crimean situation; in East Asia, we see a certain country's nuclear power development and missile launches, disputes over the Senkaku Islands in the East China Sea and the island of Takeshima in the Japan Sea, and China's determination to exploit offshore oilfields. These and other issues involve all the countries around the South China Sea, including Vietnam and the Philippines. As with the competition for petroleum and other energy resources, which has triggered world wars in the past, these tensions are also increasing the folly of war. The conflict is not limited to the region but is impacting the international situation as a whole [14]. Ian Morris states that the wave of civilizational innovation which originated in the Asian region and traveled westward to Europe, bringing with it Chinese pictographs, gunpowder, ceramic technology, and the Indian concept of 'zero', has changed direction and is heading back east toward Asia [15]. As with the civilizational progress carried by waves of technological innovation, so with the unsafety that leads from human-made disaster to natural disaster, which can probably also be transmitted along with technology.

Unsafety connected with energy resources started with the competition for access to cooking oil in the age of the great explorations and has continued through two world wars to the present day. It should not be forgotten that these conflicts that grew into world conflagrations were wars that no one thought would happen. Today, just when the new energy resource of methane hydrate has come into prospect, Russia has offered Japan a discount on natural gas and the United States has begun trade negotiations over shale gas through the Trans-Pacific Partnership (TPP). One wonders whether the world might have been peaceful if not for the trouble over energy resources. As with food self-sufficiency, an ability to meet electric and motive power needs from sustainable energy sources might have the potential to quench the fuse of international conflict. Although Japan is already equipped with a wide range of relevant technology systems, it is held back by the structural inertia associated with the politicians and research institutions (researchers and government officials) that receive donations from the electric power companies, known as the 'nuclear power tribe', plus the lobby-friendly Diet members with links to the United States, Russia, and China, and additionally the corporate groups and the old *zaibatsu* financial conglomerates which have made vast investments in the development of overseas gas resources [16]. Undeniably there is an extent to which corporate groups related to the energy business are blocking the innovation of sustainable energy, which threatens their established interests.

For instance, around the time of the Fukushima nuclear power disaster, legislative reform and deregulation were bringing a shift in Japan's renewable-energy initiatives toward solar, geothermal, and wind-based electricity generation [17], but this went into reverse with the change of government. According to NHK reporting, there are many planning applications for wind-power and solar-power generation projects which cannot be realized due to issues of approval and licensing, which means that the switch to clean energy is at a standstill [18]. A structural inertia operates through Japanese Diet members in the energy lobby to maintain established interests, allowing a switch in policy to protect the heavy investments made by corporate groups in American shale gas development and Russian natural gas.

Alongside offshore oilfields, mineral resources, and new energy resources such as shale gas, oil sand, and methane hydrate, fisheries and other marine food sources also present serious issues [19]. There are countries which do not practice stock conservation in the form of fish farming and return of eggs to the sea, but simply compete for fishery profits until resources are depleted. This constitutes a problem more serious than "Japan's whaling for scientific research purposes" and means that the topping for the sushi which is now popular worldwide may in the near future have to change from fish to vegetable. By focusing too closely on the problem of whaling, we miss the wider picture of the depletion of fishery resources as a whole. The vast majority of ordinary Japanese like myself who do not eat whale wonder who the people are who consume this strange luxury food.

Now on its way to over eight billion, the population of 'Spaceship Earth' has reached such a critical point [20] that NASA's Ames Research Center is exploring the possibility of 'terraforming' on Mars using synthetic biology (NASA AIMS) [21]. Meanwhile, the battle over water resources is intensifying. Of all the earth's water, the proportion suitable for drinking is a mere 0.008 %. With the prospect of a population explosion that will bring the world's calculated population close to 10 billion (2050) by United Nations, the struggle for water and food resources is growing serious [22], and already prices are soaring on overseas meat markets. In connection, Fig. 8.3 shows the current status of water, food, and energy resources.

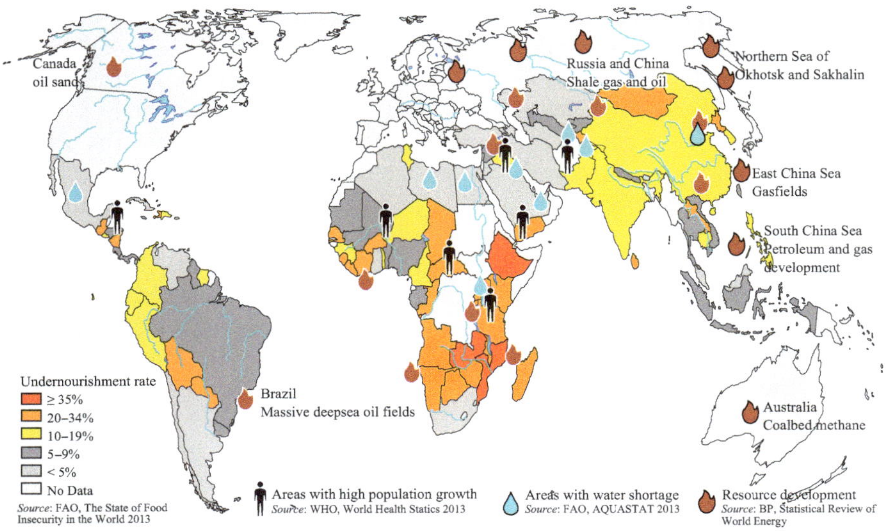

Fig. 8.3 Crises of water, food, and energy from population explosion *Note:* mapping based on inspiration from 'water management' by G. Midgley presented at ISSS Vietnam 2013. *Source data:* "The State of Food Insecurity in the World", FAO (United Nations), 2013

A natural disaster such as an earthquake, tsunami, tropical low-pressure system (typhoon, hurricane, cyclone), flood, freezing weather, drought, or water shortage (Fig. 8.1, above) may be compounded by what can be called a human-made disaster, such as war, conflict, terrorism, coups d'état, riots, or other unrest over energy or water resources, territorial rights, or religious and ethnic issues [23] (Fig. 8.2, above). Alternatively, the natural and human-made disasters may be linked, as in a food shortage arising from crop failure, competition for the protein sources of agricultural land and fishing grounds, or a struggle over the mineral and energy resources that form the economic base, and this can throw up a complex pattern that presents the characteristics of a catastrophe [24] (Fig. 8.3). Conversely, what starts out as a human-made disaster, such as the global warming from increasing levels of CO_2, chlorofluorocarbons, and methane gas, may result in the intensification of natural disasters such as super-typhoons, illustrating the undeniable possibility that human-made disaster can cause natural disaster [25]. Figure 8.4 is a graphic representation of the interconnections in contemporary unsafety, where complex catastrophe lies between human-made disaster and natural disaster [26]. 'Disaster management' includes (1) risk management (before disaster), (2) crisis management (at the time of disaster), and (3) resilience management (after disaster). The link from natural disaster to human-made disaster, and the backward reaction from human-made disaster to natural disaster, represent connections between events 'waiting to happen' whose potential to precipitate a catastrophe on the scale of Noah's Flood we must all recognize and dread [27].

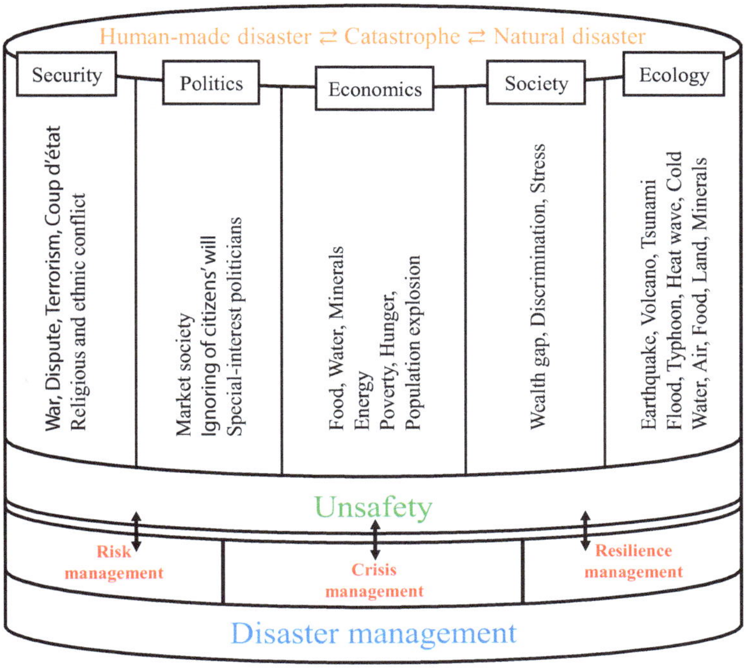

Fig. 8.4 Category of unsafety including risk, crisis, disaster, and resilience. *Note:* Drawing based on a discussion of 'disaster management' with G. Chroust at ISSS Hull 2011, Vietnam 2013 and Berlin 2015

Based on the content of the book's first half, Parts I and II (Chaps. 2, 3, 4, 5, and 6) are summarized in Table 8.1, which categorizes according to factors including disaster scale, causes, extent and type of damage, post-disaster conditions, and preventive countermeasures.

Focusing on Chap. 2, in the Fukushima nuclear disaster, the great earthquake of March 11, 2011, which reached magnitude 9.0, and the accompanying tsunami, which reached a height of 37 m, caused the six reactors at Fukushima Daiichi (Number One) Nuclear Power Plant (hereafter 'Fukushima Daiichi Plant') of Tokyo Electric Power Company (TEPCO), to cease operation. Firstly, the tsunami caused a failure of electric power supply to the water coolant for the nuclear reactors and the spent fuel pool, four of the reactors experienced meltdown and shell melt-through, and the resulting hydrogen explosion spread radioactivity; secondly there was a delay in evacuating local residents; thirdly the exhaust vents installed on government instruction worsened the leakage, but guidance to residents living 'downwinders' was neglected; [28] fourthly data from the radioactivity measurement system (SPEEDI) were kept secret; and fifthly contaminated water escaped into coastal waters [29].

Despite all of this, in April 2014, the Japanese government published a Basic Energy Plan which identifies nuclear power as an "important base-load power

Table 8.1 Case studies in this book

Case	Cause of accident or disasters	Extent of damage	Chronic/sudden Localized/extensive duration	Spread of damage Social compensation and results Cultural/social/economic/ecological damage	Preventive social function
Fukushima nuclear power plant accident (Chap. 2)	Magnitude 9.0 offshore earthquake and tsunami in Tōhoku region knock out electric power supply; control of nuclear reactors is lost	Radiation exposure of employees and residents within approx. 200 km radius (according to evacuation order of US embassy)	Sudden/extensive Semipermanent	Spread of radioactive contamination Escape of contaminated water into Pacific Destruction of communities within no-entry zone Worldwide impairment of trust	Eco-citizenship Management Policy
Environmental hormones (Chap. 3)	Impact on reproductive functions and fetal development from emission, accumulation, and biomagnification of PCBs and dioxins	Worldwide spread of PCBs and dioxins through biomagnification	Chronic/extensive	Increasingly serious environmental-hormone contamination Food insecurity Dioxins in breast milk PCB biomagnification in livestock and fish and bird species	Eco-citizenship + Management + policy
JR railway accident (Chap. 4)	Speeding train derails and overturns after failing to brake in time Background organizational factors incl. *Nikkin System* training system and use of automatic train-stop device	107 killed (incl. driver), 562 injured Loss to families of killed and injured	Chronic/localized	Compensation of bereaved not settled	Human resource management Government railway policy

Iatrogenic AIDS (Chap. 5)	HIV infection from contaminated blood preparations arising from combination of private-sector prioritization of business and sales, failings in government pharmaceutical-approval system, and bribery of academia disguised as research fees	1872 patients infected through blood products (of the total 4,028 HIV infections in Japan), 493 deaths from iatrogenic HIV infection, number of secondary infections unknown	Chronic/ extensive Whole of Japan and in some cases overseas Approx. 20 years	Secondary infection of families, children, sexual partners (number unknown). Issues of testing system for secondary infection and compensation not settled	Policy (government regulation) Management
JCO criticality accident (Chap. 6)	Shortcut in nuclear-fuel production procedure (as per secret in-house manual), labor saving in frontline operations, failure to inform staff of hazard	Exposure to neutron-beam irradiation of approx. 49 people incl. staff members, 310,000 residents evacuated	Limited area exposed to radiation Transitory	Effect on offspring of victims (contamination of reproductive cells)	Policy (government regulation) Management

source", at the very time when Japan faces the grave issue of the contaminated water still leaking from the nuclear power station. Not only the surrounding farmland and the inshore fishery but local communities and local industry have been impacted, so that society as a whole, along with export trade, has suffered damage, making this a major disaster with permanent future consequences [30]. Local communities have collapsed, farming and fishing families have been deprived of their livelihood, and the living environment and life has itself been threatened by radioactivity and contaminated water [31]. The nuclear disaster thus violated constitutional law, penal law, and civil law [32]. Perhaps the greatest failure to understand its gravity is displayed by the perpetrators of the accident, the power company overseeing the electricity facilities including the Fukushima Daiichi Plant, which is seeking to use public finances to fund compensation [33].

1. At first, the accident at the Fukushima Daiichi Plant was seen as a nuclear disaster brought on by the earthquake and tsunami natural disasters, but later its character, to some extent a human-made disaster, was revealed by the results of inquiries by the National Diet Accident Investigation Commission (NAIIC, 2013) [34], and the Civil Accident Investigation Commission (Civil Accident Investigation, 2012). Firstly, there was the irrational location of the nuclear power station. Here we can point to the irrationality of locating ten nuclear reactors in an earthquake-prone area, which was subject to four earthquakes of intensity 5 or above on the Japanese scale in 2010, a plan carried out by in part forcibly suppressing the opposition of local residents. Whatever the convenience of locating electric power supply close to Tokyo, locating a nuclear power facility here meant concentrating 10 nuclear reactors, the combined total of the two Fukushima nuclear power plants in a high-risk earthquake-prone area.
2. Secondly there were issues in the safety-management system of TEPCO. At the Fukushima Daiichi Plant facilities, there had been 120 earlier technical incidents: 'cracks in the pressure wall shroud' had been underreported as 'loosening of bolts' and recorded data on accidents and technical incidents had been falsified and systematically covered up. These facts were made public in submissions to the National Diet by the Ministry of Economy, Trade and Industry and the Nuclear and Industrial Safety Agency. The possibility that the disaster at the Fukushima Daiichi Plant was an example of Reason's 'organizational accident' [35] which was waiting to happen was pointed out by the National Diet Accident Investigation and the Civil Accident Investigation Commissions.
3. Thirdly, despite the repeated technical incidents at the Fukushima Daiichi Plant, the company had ignored its obligation to report them to the regulatory authority, the Nuclear and Industrial Safety Agency. Questions were raised by the Citizens' Nuclear Information Center over this and other systematic irregularities in safety control which led to the systematic cover-up of incidents and the falsification of data [36].
4. Fourthly, there was an issue regarding the supervisory responsibilities of the Japanese government, with administrative authority replicated by a number of regulatory government offices that together bore responsibility for approval and

licensing. These were notably the Nuclear and Industrial Safety Agency (part of the Ministry of Economy, Trade and Industry) and the Science and Technology Agency, which were responsible for nuclear power policy, but there were also for instance, the Nuclear Safety Commission and the Japan Nuclear Energy Safety Organization. The official regulatory system for nuclear power was duplicated and triplicated, leading to a kind of systems pollution [37]. Moreover, as a policy to 'soften up' the local community and stakeholders [38], public funds were distributed to silence the opposition on the basis of bad laws (the *Three Power Source Development Laws*) mandating payments in return for exploitation of resources. These encouraged a baleful structural inertia in local communities, municipal authorities, and local industry and reinforced the phenomenon of 'unstoppable nuclear power'.

5. Combining with this structural inertia, it seems fair to say that there was a collapse not only of technology functions but also of business executive functions as the maintenance of the aging nuclear reactors' facilities exceeded the limitations of management. It seems clear that the aging nuclear power station had a crushing effect on the workplace staff, who adopted a "see no evil, hear no evil" attitude. All three groups—operational managers, business executives, and technicians—seem to have regarded it as a power source that "made money but was always breaking down", as many nuclear power technicians and workers have testified. The Ministry of Economy, Trade and Industry and the Science and Technology Agency, which were the regulatory government authorities, allowed the power company to deliver its deceitful reports in written form only and failed to carry out on-site inspections. Instead of closing down the aging nuclear power station, they allowed it to carry on operating like a toy that makes money by spewing out radiation.

8.2 Socio-homeostasis: From Disaster Management to Crisis Sciences

To summarize, the Fukushima Daiichi Plant suffered from a number of problems, notably (a) irrational location and incomplete technology; (b) cover-up of nuclear power technical incidents, falsification of data, and management failures in the area of safety control; and (c) ambiguous nuclear power regulation arising from the web of vested interests surrounding approval and licensing operations. This was the power station that was struck by the earthquake and tsunami: the world's weakest facility of its kind, the Fukushima Daiichi Plant with a 'systems pathology' [39] in terms of (a) physical location (earthquake-prone area), (b) inadequate safety control within the organization, and (c) ambiguous regulatory supervision. Ironically, just 1 month before the Fukushima nuclear disaster, the proposal had been made to extend the operating period of Japanese nuclear power plants from 40 years to 60 years, which may well be considered surprising (Nuclear Safety Commission, 2010) [40]. Moreover, in October 2010, less than a year earlier, the Japan Nuclear Energy Safety

Organization had reported that, in the event of electric power loss at the Fukushima Daiichi Plant's nuclear reactor, shell melt-through would take place in one hour and forty minutes (Japan Nuclear Energy Safety Organization, 2010) [41]. But these alarm signals were ignored, and the nuclear accident feared by everyone including the company, the government, and local residents—to put it simply—waited no longer to happen [42].

Historically, major accidents and disasters appear to happen against a background of systemic collusion and structural inertia in business management and public policy. Failure to prevent this kind of structural inertia in the social organization invites major accidents and disasters. In April 2014, there was a major shipping accident in South Korea when freight overloading caused the ferry *Sewol* to capsize. Mishandling of the evacuation led to the tragic loss of many lives. Unfortunately for South Korea, the country was hit around the same time by a series of incidents, including a subway accident and fraud involving aircraft and nuclear power components, exposing a collusive structure between financial and political interests. These networks of corruption in businesses and politics result in major disasters and accidents and stir up social anxiety [43]. As mentioned earlier, authoritarian or religious states which wear the mask of a constitutional state have a different social ethos and cultural background to that which has been fostered by the history of the developed countries. However, irrespective of cultural and ethnic disparities, it remains true that structural inertia in the social organization results in major misfortune. In whichever country, east or west, regardless of religion or race, and transcending differences of social system, preventing unsafety is an urgent task for modern society. For the sake of a small number of wealthy individuals, it should not be necessary for the rest of the public to bear the risk associated with accidents and disaster.

As a social function to prevent unsafety, I would like to present the ideas of some progressive thinkers. Regarding the preventive function in society, one may suggest the idea of a 'social morality' which integrates and balances organizational effectiveness and individual efficiency [44]. According to C. I. Barnard, the benefits of cooperation lie in monetary earnings and other elements of economic value, but also in a sense of belonging, social welfare, medical treatment, healthcare, pensions and other aspects of social surplus. At the same time, the economic profit and social trust brought to the organization as a whole by cooperation raises organizational effectiveness [45]. This is reflected in the organization's ability to raise funds on the financial and bond markets and also in its internal reserve. In addition to this binary system of metrics for the organization and the individual, the author advocates the third metric of social morality. Thus, to supplement 'efficiency'; the index of the individual (X axis), and 'effectiveness'; the index of the organizational total action (Y axis), he posits a measurement relative to society, which is 'morality' (Z axis), and recognizes the harmonization and integration of the three as the role of management (executive function and responsibility) [46]. By raising social morality as the third dimension alongside organizational effectiveness and individual efficiency, and adding the fourth-dimensional time axis to stand for continuity, Fig. 8.5 represents sustainability as a preventive social function.

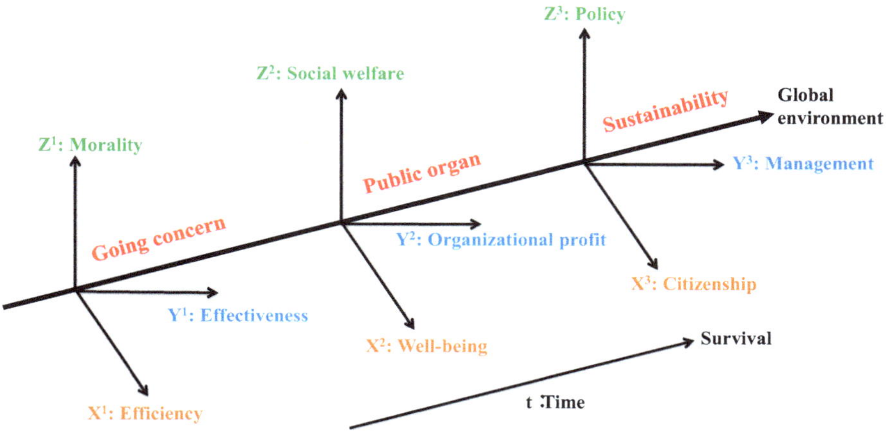

Fig. 8.5 4D Sustainability as a social function

In the advanced nations and in developing countries also, economic growth is founded on economic rationality and social rationality. H. A. Simon calls the rationality at the base of modern society 'bounded rationality' [47]. It consists of the following three aspects:

1. Human multiplicity: the multiplicity arising from the synchronistic effect of 'action and intellect' in individual behavior;
2. Diversity of action: the diversity of nonrational action and mutual reaction arising from complex human behavior;
3. Relativity of observation: the history of the irrational results of relative action in human cooperation and the relativity of observation.

What is accepted as reason in a certain age under certain conditions is often seen as 'unreason or irrational' in a different age. On this point, Barnard focuses on the idea of latent rational control in human action and suggests a latent 'nonlogical process' [48]. He foresaw that the 'unsought consequences' that are the result of human cooperation not only could endanger organizational survival but also might inflict damage on society. Writing in a similar vein, Drucker speaks of 'unexpected results' [49]. At the root of rationality is latent human desire, and there can be no such thing as absolute rationality, which is no more than a fiction applying in a relative sense within a limited sphere (age, society, religion, or culture). Essentially, rationality involves the 'common sense' of people determined by the times they live in and their 'psychological situation', and it is their common consciousness, its conventions and codes, that form social rationality [50]. Incidentally, we might say that nature moves in a way that transcends the rationally based ideas of humankind; having progressed from absolutism (monarchy) through feudalism, commercialism, socialism, and communism to capitalism, global rationality is now shifting toward sustainism [51].

Human rationality itself arises only out of a limited set of physical, biological, psychological, social, and historical conditions and, adapted to and contingent on our total situation, has no more than the outward appearance of a measurement standard. The standard of rationality is subject to religious, cultural, and linguistic differences, for instance, between Christians and Muslims, and is an essentially limited 'relative standard'. G. E. Swanson suggests that even scientific rationality is controlled by a language scale and a mathematical scale, and speaks of the limited nature of the social mission of scientists [52]. As the same, social rationality is different from scientific rationality by G. A. Swanson.

Meanwhile, H. R. Maturana, speaking of the relationship between reason and emotion, says that reason and its counterpart, ethics, are the 'translational systems' or form of programming "founded on love and emotion and serving the purpose of the conservation and change of the species" [53]. Human behavior is regulated not only by linguistic knowledge; the domain of tacit knowledge, involved in empirical knowledge and behavioral knowledge, is also in effect [54]. The resulting 'empathetic majority' is a universal phenomenon irrespective of race, ethnicity, culture, or language and is rooted in the instinct for survival of the species shared by all human groups as biological organisms [55]. This means that humankind's universal instinct and emotion for survival are the foundation of ethics. This 'empathetic collective consciousness' which people share, seen in another light, is citizen power. Standing in opposition to the social functions of business management and public policy, which are aggrandized by capitalism and democracy, it represents another social function capable of realizing sustainability for the community. The collective consciousness of the majority of citizens is what forms modern rationality and is speaking up for a global ecosystem permitting human survival. Sustainability to overcome the limits of capitalism, which prioritizes the logic of the rich, is the clarion call of today [56].

During Japan's postwar era of rapid economic growth, the prioritization of material wealth received the 'empathy' of the Japanese population of the time. It was an age when a universal tacit agreement had formed within the social organization over the 'logic of economic prioritization.' Not only in Japan but in almost all countries enjoying periods of rapid economic growth, this growth has been given precedence despite being the cause of industrial effluent, atmospheric pollution through exhaust gas, and soil contamination [57]. The upshot of rationality is seen in the government officials and politicians who have lost the 'human touch', who may have ability but are lacking in humanity, whose decision-making and judgments are dominated by the power of established interests, and who turn away from the laments of the people and pretend not to see. The history of dealing in this way with public grievances is not unique to Japan. In those days, the age of the 'convoy system' and the mocking phrase *Japan Inc.*, economic growth carried the majority, and the *Japanese-style rationality* which took precedence expressed the popular will in support of high-rate economic growth. A similar approach can be seen with the United States, which plays the lead part in the money game, and the BRIC countries (China, Russia, India, and Brazil), which are riding on a wave of economic success. Within the context of the world market, countries such as these

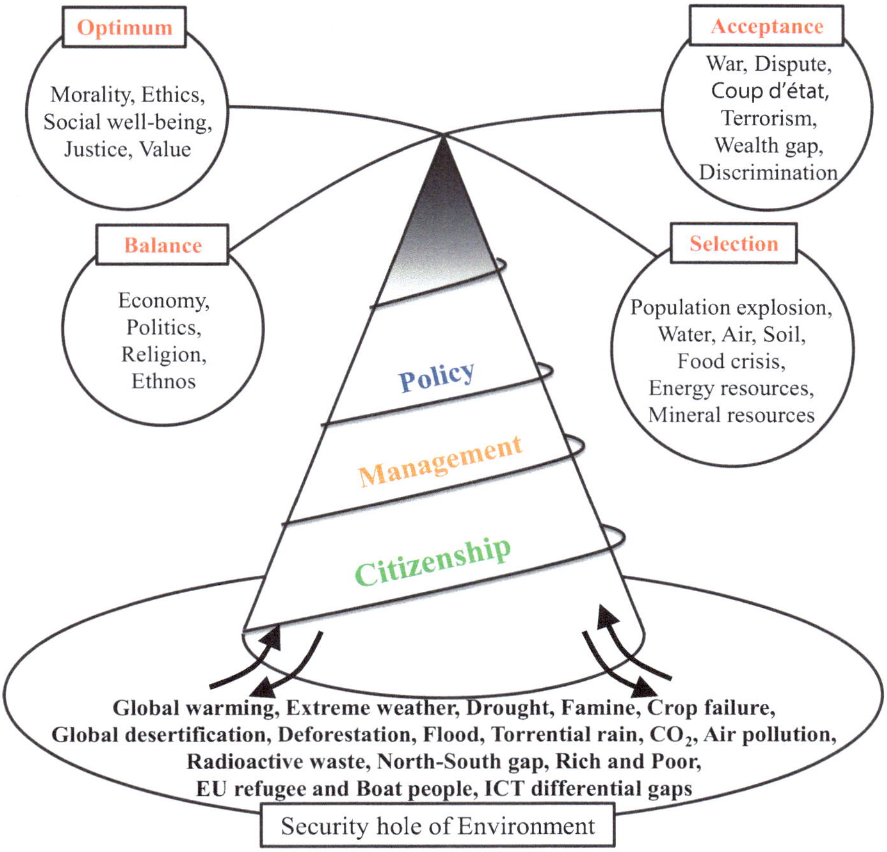

Fig. 8.6 Socio-homeostasis: management, policy, and citizenship

ignore international rules on intellectual property rights and territorial rights, while certain countries manipulate the exchange rate of their currencies as part of state policy. Here lie the basics of the world economy and the sources of social anxiety.

Going forward, citizen consciousness is shifting from the age of growth to the age of sustainability, and the consensus around citizenship, which embodies the people's shared consciousness, will determine the rationality of the near future. A theme debated at the 2013 conference of the International Society for the Systems Sciences in Haiphong, Vietnam, was the incipient idea of an *eco-civilization* based on citizen power to oppose business management and public policy with their tendency to fall into collusive relationships [58]. Figure 8.6 shows a representation of the role of 'eco-citizenship' in preventing impairment of social functions by business management and public policy.

To review the examples of unsafety examined herein, the book started with the Fukushima nuclear catastrophe in Chap. 2 and followed with the 'biomagnification' of the environmental hormones PCB, dioxin, and other chemical contaminants [59] in Chap. 3, the accident on a Japanese railway that prioritized commercial profit in Chap. 4, the disaster of iatrogenic HIV infection with its 100,000 plus victims [60] in Chap. 5, and the JCO criticality accident, which exposed the local community to radiation, in Chap. 6. Common to all of these is the downside of the organization, whose attachment to commercial profit and economic prioritization results in an antisocial lack of humane consciousness. These accidents and disasters not only deprive us of life, health and property, but also inflict damage on the ecological and social environment to the detriment of the next generation. Organizations that cause accidents and disasters no doubt pursue profit on the basis of economic rationality, but at a stroke the accident leaves them with vast compensation problems and the loss of public trust for generations. Having prioritized economic value and efficiency in defiance of public sentiment, business management and public policy end up causing the accidents that were waiting to happen, and find it difficult to restore the lost trust. The accumulation of unsafety is making our stolen future irrecoverable. Modern science, in order to prevent these great tragedies, needs to restructure the compartmentalized and micro-divided world of individual scientific disciplines. In connection, G. Bateson discusses the danger of science becoming individualized, subdivided, and compartmentalized [61].

In a new approach that draws on the tragic lessons of ferry accidents in 1987 and 1989, Britain established the concept of 'corporate manslaughter' in social organizations [62]. The legislation introduced, which examines the criminal liability of the corporate organization in cases of organizational accident or disaster, has had a preventive effect. The social concept of corporate manslaughter, which was formulated in line with the sympathies of the British people, has been effective as a prescription to prevent great misfortune. The mechanism of the social function to prevent unsafety, including the ***risk–crisis–resilience management*** for humanity [63], is represented in Fig. 8.7.

In Western thought, Aristotle suggests that 'policy and ethics' are two sides of the same coin, and that the consciousness shared by individuals promotes the formation of consensus and thereby determines policy. Aristotle claimed that underlying politics is the fact that ethics is latent in the people's common consciousness, such that policy is the expression of ethics. In Eastern thought on the other hand, T. Watsuji, in his work (1979), writes that ethics is in all aspects expectant on politics, which in a sense presupposes ethics; the two are thus in a mutually complementary relationship [64]. This complementarity of policy and ethics allows them to adjust at a preconscious level to the changing trends in the diverse values that mediate between the individual and society in order to realize the 'good'. Ethics, which is at the deepest level of people's spiritual foundation, draws on the innate good within humans as its motive force. Policy is the social function that actuates the ethics contained in this tacit aspect of the common consciousness, and the collective consciousness that informs policy is a composite construct which can be called ethics as 'social convention'.

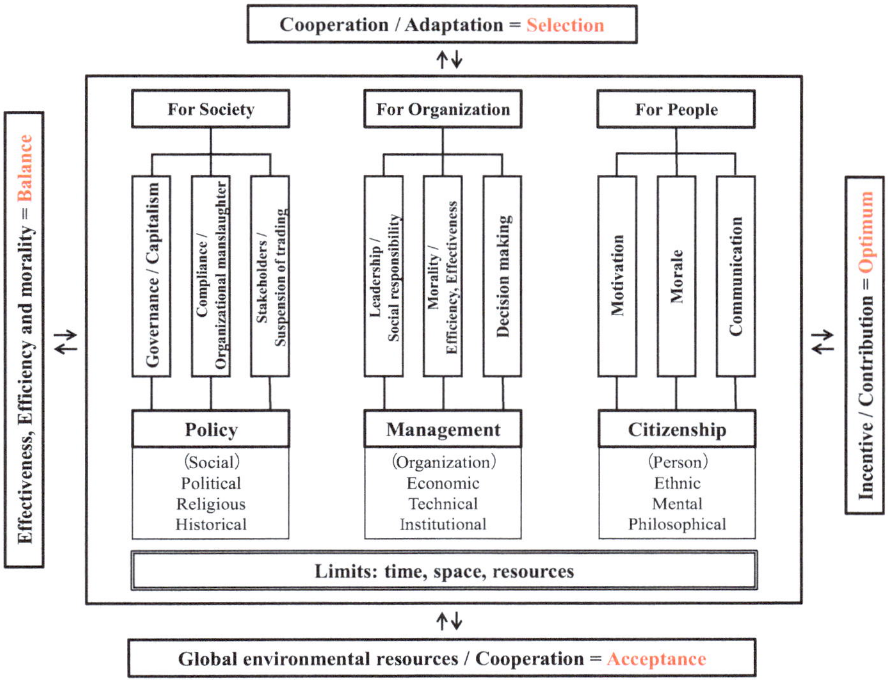

Fig. 8.7 Functions of disaster managements for people, organization, and society. *Note:* Drawing based on discussion with M. Morley's presidential speech 'Global HRM' at IFSAM Ireland 2012 and Tokyo 2014

M. Sandel views as problematic the development "from the market economy into the market society", which he says has brought us to 'moral limits and dilemmas' [65]. The point of concern is that the market principles of modern capitalism have altered the nature of society, such that the market, which is a device to drive economic action, has at some point become integrated into the structure of society and changed it into a market society. This marketization, creating a society which ignores social well-being and decides all in terms of the market, is a driver of inequality. In response, Sandel points to the 'moral dilemmas' of extinguishing society, culture, ethnic identity and religion, and destroying the wisdom and conscience. From this philosophical standpoint, he warns of the risk of undermining the foundations of democracy along with freedom, equality, and human rights.

8.3 Crisis Sciences for Our Survivability: Eco-civilization

The very concept of the environment is 'translational systems', from which it is difficult to derive a fixed concept. The existence of the environment reflects von Bertalanffy's 'system thinking' in some cases, but it exists as a 'chaos and fractal' phenomenon in other cases [66]. Environment as a global ecosystem is the field of human symbiosis, and its time-space transcends human wisdom. Both the ecological environment perceptible as a physical phenomenon and the social environment in the ecosystem which is difficult to visualize have microscopic and macroscopic aspects, so that regardless of the distinction between organic and inorganic, the dynamic condition is the reality. Also relevant to the social environment are not only galactic space, solar activity, the moon's tidal force and other spatial, physical, chemical and biological factors [67] whose effect can be seen in global history, but also social, economic, political, ethnic, cultural, and historical factors originating from human activity. However, the long-term benefits of human cooperation do not have an exclusively positive effect but reshape the global environment. The cumulative negative effects of this have led to the ozone hole caused by CFC gases, what are at the same time greenhouse gases, as well as the unsought consequences of CO_2 and methane gas [68]. Sum of this accumulated unsafety brings on natural disasters and is the origin of great tragedy. But who will speak in defense of the environment? Should it be politicians; should it be business managers? But can the politicians and businesspeople, with their tendency to collude in defending their interests, be trusted to speak up for the ecological and social environment?

Apart from the environment, whether as chaos or fractal [69], physical, biological, and social elements are involved, and historical, cultural, civilizational, and psychological aspects are latent (Y. Shiozawa) [70]. For these reasons, understanding of the environment proceeds through both empirical interpretation and linguistic interpretation, and it exists more strongly as an imaginative idea than as an intellectual concept. Regarding perceptions of the environment, even where the words used are the same, the content will differ according to country and generation, professional standpoint and sense of values, lifestyle, wealth differences, and other factors. The environment is known as a system integrated with its own process [71], and exists as the temporal-spatial, historical, philosophical, cultural, and social totality of the global ecosystem [72]. For instance, perceptions of the environment among workers in agriculture, fishery, and other primary industries will sometimes be completely opposite to those of executives in the secondary industry of manufacturing. For business managers, the environment is something to be processed, but for agriculture and fishery workers, the current environment must be preserved for their harvest. Capital, strengthened by the stock market, has made use of mineral resources and energy to practice mass production, mass consumption, and mass disposal [73]. Unlike farming and fishing folk, who wish to preserve the environment, enterprises in mining and manufacturing process the environment to create new wealth. The various technologies that support civilization have been utilized to this end, but with the unsought consequence of environmental destruction, a negative aspect which has come to be seen as problematic today [74].

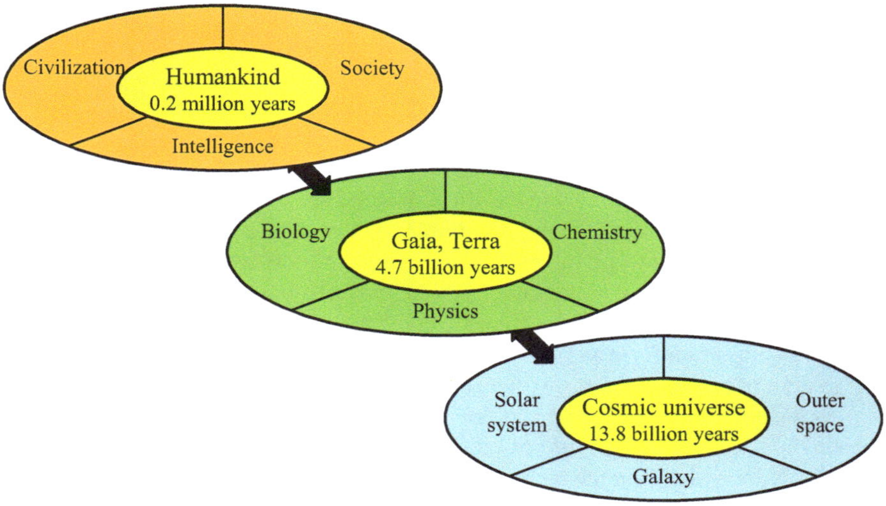

Fig. 8.8 3D mandala of Gaia's unsafety

Ideas on how to prevent the civilizational collapse arising from this environmental destruction cannot be gleaned from space aliens or the people of the future. Nor can the answers be found by following through the technological systems that have brought environmental destruction in their wake. The time has come to rethink and move on from the conventional progressive view of history toward a sustainable 'Gaia-based view of history.' As we cannot expect to be taught the art of realizing global sustainability by 'gods', we can only look to past wisdom, where we may pick up some clues from the fact of humankind's 200,000 years of existence. The sustainability against crises to which modern society aspires can be found by tracing back to the ecological wisdom of the native peoples who have survived to the present day (Fig. 8.8).

According to Y. Tsukio (2012), in his book *Journeys Beyond Time and Space: The Wisdom of the Indigenous Peoples* [75], presented ancestor's intellectual wisdom as a prototype of 'eco-civilization'. For instance, the native American Iroquois tribe made their "decisions after thinking of their descendants down to the seventh generation." For the sake of their descendants in future generations, they practiced decision-making to leave a livable environment, thus making sustainability an actual part of tribal custom. This 'back-cast' approach is a method that envisages the ideal future, then reckons backward from the future to work out the ideal way for the present. Seen through the lens of this back-casting, modern capitalism is a social system that consumes mineral and energy resources rapidly and places central emphasis on affluence and convenience, which can only be described as an approach that imperils the preservation of the species for future generations [76]. The Nobel Peace Prize winner and native Guatemalan Rigoberta Menchú says that the natural wilderness is a more important resource than the petroleum, mineral, and other resources sought after by greedy people; and that humans have learned the spirit and life of sustainability from mother Earth, the sea, the sun and the moon over many generations, and must abide by these lessons for the sake of future generations (R. Menchú). Sustainism was also reflected in the hunter-gatherer economy of Japan's 'Jōmon culture', which endured for 12,000 years.

The Maori people native to New Zealand hold nature and the forest as sacred and originally thought of the land and the water as goods to be held in common, having no concept of ownership. Under the capitalist system, in contrast, land ownership and water resources are privatized, creating a social structure which views it as unproblematic to excavate land under private ownership, or release exhaust gas or wastewater there. For the native peoples, however, the land could not be divided up, and it was important for water resources and the air not to be in private ownership. The system of private property with its concept of individual ownership was located in their zone of indifference and was not understood or recognized. There had never been such an idea. Chief Seattle of the Native American Seattle tribe said that the idea of buying and selling the blue sky and the open country was incomprehensible. The land and the air were of their very nature common goods, and this social convention remains part of the tribe's social intelligence regarding the ecosystem (Chief Seattle). The social conventions of these groups of native peoples, reflecting their environmental survival wisdom, are graphically represented in Fig. 8.9, categorized into the frameworks of the previously mentioned linguistic knowledge, behavioral knowledge, and empirical knowledge [77].

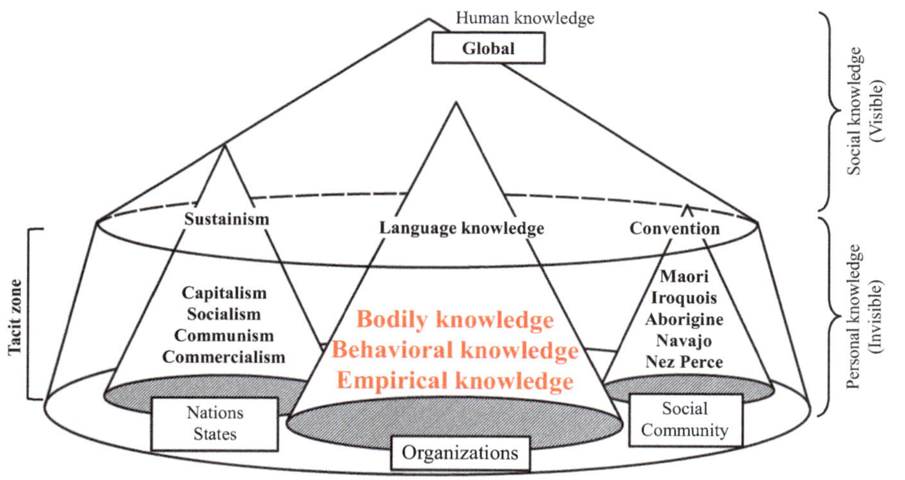

Fig. 8.9 Eco-civilization as social intelligence and wisdom. *Note:* Inspired from H.R. Maturana "Ethics as a Human Action" (ISSS in Toronto, 2000), M. Polanyi *The Tacit Dimension* and A.J. Ayer *The Foundations of Empirical Knowledge*. *Source*: Atsuji S., *Soshiki Kettei no Jyouhou Katei* [*The Informatics of Organizational Decision-making*], Japan Academy of Business Administration, 2000, pp. 128–135

The idea that today's intact natural environment is entrusted to the present by our future descendants (Navajo tribe) cannot be said to have even been considered in the developed countries that operate under contemporary capitalism, which exploits resources to the maximum during periods of economic growth. This illustrates a blind spot of the modern capitalist system as it seeks commercial profit for economic effect. What we can learn from native peoples is that we may be entering a period of paradigm shift from the "limits of capitalism to sustainism." Looking back over the history of the native peoples, we see that they have survived for tens of thousands of years (I. Morris), in contrast to the modern social system, which has lasted barely 200 years and today displays wide-ranging evidence of decay. Common to every native people that has achieved long-term survival is a system of social convention to support sustainability and clearly established rules for the maintenance of 'symbiotic civilization' that respect the ecosystem and are carried on by successive generations based on the social brain as 'global brain' [78]. The conceptual framework of crisis sciences [79] based on eco-civilization as the intelligence of social wisdom is represented graphically in Fig. 8.10.

The cause of the world wars that humankind has experienced is competition over petroleum, coal, and other energy resources. The offshore oilfields and methane hydrate of the South and East China Sea and other energy resources have led to heightened tensions in East Asia. In response to this energy problem, a method based on a 'self-contained' natural energy system for mountainous regions is likely to set an example for renewable energy of the near future. In the kingdom of

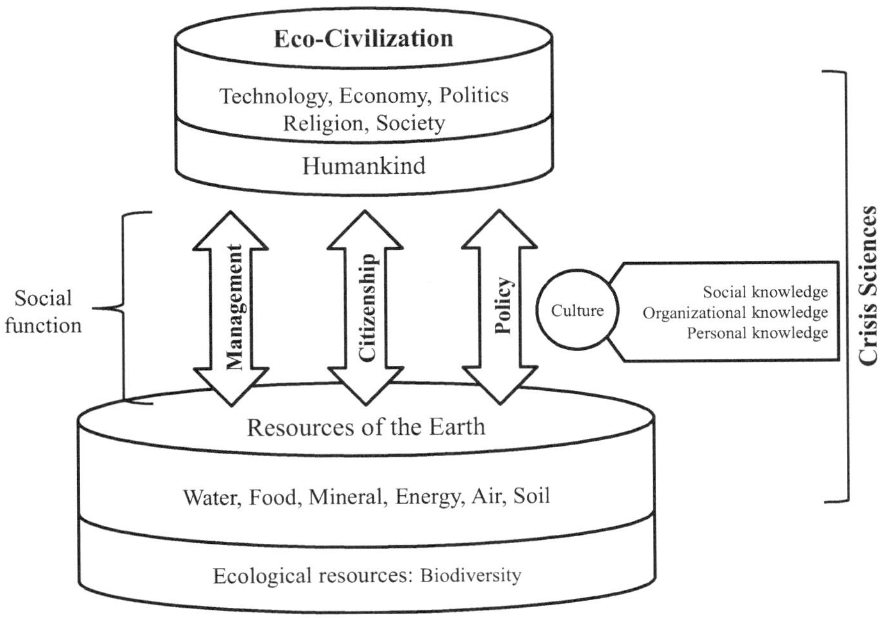

Fig. 8.10 Framework of 'crisis sciences' for eco-civilization

Bhutan, springwater from steep mountainsides used to drive small hydroelectric plants is being adopted as a decentralized form of renewable energy operating at the village level. Sustainability, rather than being a civilizational device of the capitalist system serving an immediate purpose, is an essential human approach considerate of family and future generations. According to the fourth Wangchuck King of Bhutan, this development means reclaiming a uniquely human spiritual and cultural adaptation to eco-civilization (Wangchuck). This contrasts with the increasingly unsustainable expanded 'wealth gaps and imbalances' represented in the world's largest economy [80], the United States, and the escape from reforming social insurance and medical treatment insurance systems attempted by President Obama. Are the people of the world's most affluent country really happy? (*See*, e.g., Detroit City problems.)

According to Professor Tsukio, forest accounts for only 12 % of the land area in Britain, but 68 % in Japan, where the natural ecosystem was apparently preserved by allowing local shrines to draw boundaries between 'home forest and deep forest'. J. Diamond states that the Edo Shogunate established strict rules on the felling of forest, which was both an energy resource and a timber source, thus managing and protecting forest resources [81]. World's four great civilizations grew up in the basins of great rivers. In the kingdom of ancient Egypt, where to rule the water was to rule the land, being able to predict the flooding of the River Nile plain from the movements of the stars and the sun gave control over fertile lands and agricultural produce. Nowadays, dam building has deprived the River Nile basin of some of its agricultural land and caused crop failure [82].

The conclusion to draw from these considerations is the clear principle that humans are not in opposition to nature but are part of the environment. The idea of eco-civilization remains alive today in the worldwide spiritual ethos and in the implicit convention that natural law and human practice must be harmonized. Crisis sciences are the wisdom inherited from the previous generation to at least guard against 'unsustainability'. The content outlined above is presented below in broad outline in graphic form as an 'unsafety tree' (Fig. 8.11). It may be noticed ironically that the tree depicting the natural disasters and human-made disasters that threaten human survival is analogous to a tree depicting scientific disciplines. Thus, the science of crises has prevented unsafety and there is an intelligence of eco-civilization based on human wisdom. This illustrates that our history and the global environment which integrate it are two sides of the same coin linked by human cooperation.

8.4 In Conclusion

In prehistory, while hunting the mammoth and the mastodon as a food source, humankind extended its range from Africa to Europe and Eurasia, and then to the American continent [83]. With one mammoth, a family had enough food and hide to keep hunger and cold at bay for half a year, and its bones could be used to build a portable shelter. In this way, humankind, amid the Ice Age, was able to extend its

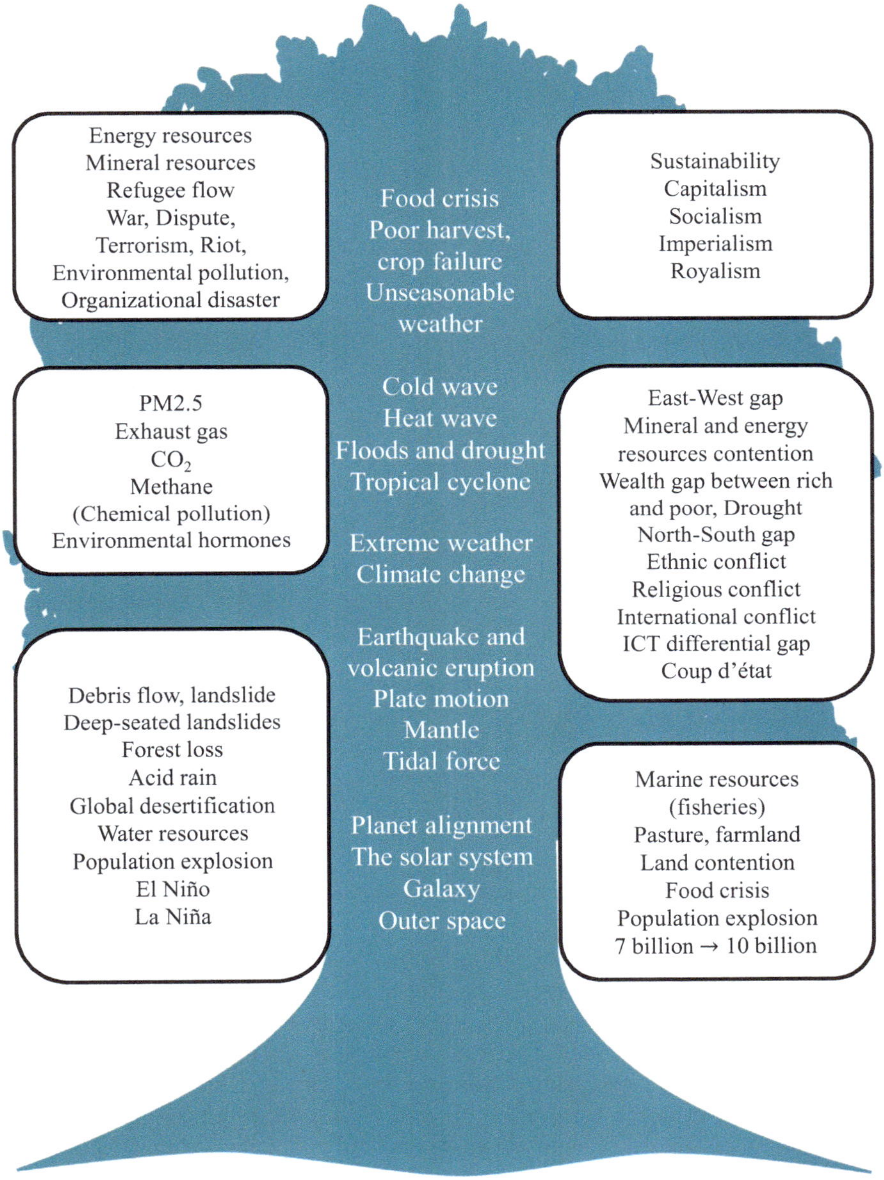

Fig. 8.11 Unsafety tree of crises taxonomy

domain to the whole of the earth. But having eaten the mammoth to extinction, humankind lost its food source, and had to overcome a very challenging period by making the transition from hunting to agriculture. Thanks to the forest, there is a place to live in shelter from the heat and the cold. The forest provides us with a

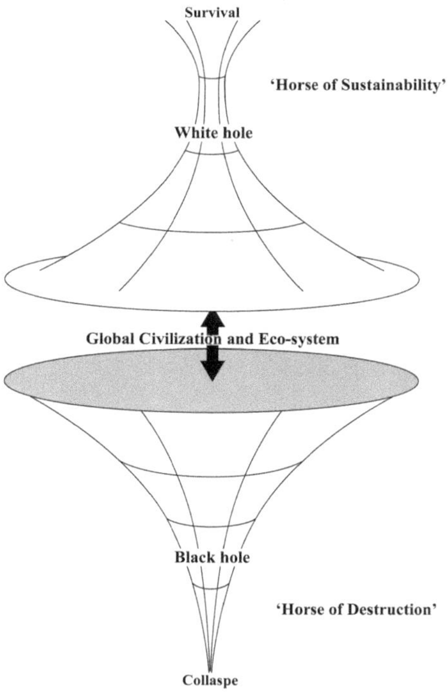

Fig. 8.12 Diehard decision-making for survivability. *Note:* Analogy based on black/white holes from 'Horse of Sustainability vs. Horse of Destruction' by J. Diamond

water supply by retaining moisture in hillsides and feeding it into rivers. Without the forest there would be no human life. The forest fosters animal and plant biodiversity, and at the same time has been a shelter and food source in the ecosystem. It may be an irrelevant fact, but it has been estimated that if the deserts of Africa, America and Eurasia could be greened to become like Siberia or the Amazon Forest, humankind would be three to four times richer.

As part of daily life, humankind has faced the vicissitudes of the natural and human-made disasters occurring in the global ecosphere of Terra (Fig. 8.12). In people's everyday life, the important issues cover everything from crop failures caused by abnormal weather patterns to human relations in the workplace, death and suicide from overwork, and traffic congestion. To take a macroscopic view of human history, the earth is now in an interglacial period between ice ages. Whether the cause is CO_2-based human-made disaster or alignment of the planets and supernova explosion, we are said to be entering a period of great climate change, in which earthquake, tsunami, typhoon, flood, extremes of cold, drought, and other events will be crucial issues for human survival. Regardless of whether the CO_2 emitted by human civilization is the cause of global warming—which in any case is a microscale phenomenon in the context of Gaia's cosmic evolution over 4.7 billion years as part of the solar system and the Galaxy—humankind's wish to continue living on this earth remains the same. There may be climate crisis due to cosmic radioactivity released by supernova explosions in extragalactic space; there may be activation of gravity and mantle-plate convections due to the 2012 planetary alignment in the solar system; there may be war and conflict. Nevertheless, it is human nature to want to carry on living. This means that the time has come to respond to the natural disasters that strike humankind, the unsafety that we bring on ourselves and the catastrophes that arise from the combination of the two aspects by making them the subject of scientific study. The scientific study of crises is quite simply the mission of eco-civilization for humankind, "sustainable decision-making" for ecological survivability.

References

1. Diamond, J.M., *Collapse: How Societies Choose to Fail or Succeed*, Viking Penguin, 2005.
2. Diamond, J.M., *Guns, Germs, And Steel: The Fates of Human Societies*, W. W. Norton & Company, 1997.
3. Brown, L.R. et al., *Saving the Planet: How to Shape an Environmentally Sustainable Global Economy*, Worldwatch Institute, 1991.
4. Atsuji et al. "The Fukushima Catastrophe seen as a Malfunction of Organizational Systems: Nuclear Fade-out and Hazard Maps for the World", *General Systems Bulletin*, Vol. XXXXI, 2012.
5. Beck, U., *Risk Society: Towards a New Modernity*. London: Sage, 1992.
6. Brown, L.R., Flavin, C. and Kane, H., *Vital Signs: The Trends that Are Shaping Our Future*, Worldwatch Institute, 1992.
7. Brown, L.R. et al., *State of the World 2000*, W.W. Norton, 2000.
8. Cribb, J., *The Coming Famine: The Global Food Crisis and What We Can Do to Avoid It*, University of California Press, 2010.
9. Birkmann, J. (ed), *Measuring Vulnerability to Natural Hazards*. New York: United Nations University Press, 2006.
10. Sachs, J.D., *The End of Poverty: How We Can Make It Happen in Our Lifetime*, The Wylie Agency, 2005.
11. Acemoglu, D. and Robinson, A., *Why Nations Fail: The Origin of Power, Prosperity and Poverty*, Brockman, Inc., 2012.
12. Beck, U., *World at Risk*. Cambridge: Polit, 2009.
13. The World Bank, *Natural Disaster, Non-natural Disaster*, 2011, p.30. *Available* at: https://www.gfdrr.org/nhud-home. *Accessed* May, 2014.
14. Meadows, D.H., Randers, J. and Meadows, D.L, *Limits to Growth: The 30-Year Update*, Chelsea Green Publishing, 2004.
15. Morris, I., *Why the West Rules--for Now: The Patterns of History, and What They Reveal About the Future*, Picador; 2011.
16. McMillan, C.J., *The Japanese Industrial System*, De Gruyter, 1996.
17. Yergin, D., *The Quest*, William Morris, 2011.
18. *See*, NHK ONLINE. *Available* at: http://www.nhk.or.jp/gendai/kiroku/detail_3471.html#marugotocheck. *Accessed* May, 2014.
19. Brown, L.R., *World on the Edge*, Earth Policy Institute, 2011.
20. Fuller, R.B., *Operating Manual for Spaceship Earth*, Southern Illinois University Press, 1969.
21. *See*, NASA AIMS. *Available* at: http://www.nasa.gov/centers/ames/home/. *Accessed* April, 2014.
22. Smith, L. C., *The World in 2050: Four Forces Shaping Civilization's Northern Future*, 2010.
23. Brown, L.R., *Plan B 4.0: Mobilizing to Save Civilization*, W.W. Norton & Company, 2009.

24. Wisner, B., Blaikie, P., Cannon, T. and Davis, I., *At Risk: Natural Hazards, People's Vulnerability, and Disasters (Second Edition).* London: Routledge, 2004.
25. Turner, B., *Man-made Disaster,* London: Wykeham, 1978.
26. Meadows, D.H., Meadows, L.M., and Randers, J., *Beyond the Limits*, Chelsea Green, 1992.
27. Thom, R., Zeeman, E.C., Ushiki, S., Sawa, T., *Keitai to kōzō—katasutrofu no riron* [*Catastrophe Theory*], Misuzu Shobō, 1977.
28. *See*, NHK documentary wave. *Available* at: http://www.nhk.or.jp/documentary/. *Accessed* April, 2014.
29. Atsuji S. and Cook, N.D., "Fukushima Nuclear Catastrophe 3.11: System Pathology of Social Organizations", *ISSS*, 2011.
30. Atsuji S. and Ueda, K., "Un-safety: Systems Pathology of the Fukushima Nuclear Catastrophe", *ISSS Proceedings*, 2013.
31. Atsuji S. and Ueda, K., "Sustainable Decision-Making Following the Fukushima Nuclear Catastrophe", *IFSAM*, 2012.
32. *See*, News for the People in Japan. *Available* at: http://www.news-pj.net/diary/1001. *Accessed* May, 2014.
33. Atsuji S. and Senba, S., "Denkijigyōjiyūka to shakaikōkyōsei", in *Kōgyō keiei kenkyū* [*Industrial Management Research*], Kōgyō Keiei Kenkyū Gakkai, 2002, vol. 16, pp. 22–26.
34. *See*, The National Diet of Japan, *The official report of The Fukushima Nuclear Accident Independent Investigation Commission*, 2012. *Available* at: http://warp.da.ndl.go.jp/info:ndljp/pid/3856371/naiic.go.jp/en/report/. *Accessed* May, 2014. *See also*, Civil Accident Investigation Committee.
35. Reason, J., *Managing the Risks of Organizational Accidents*, Ashgate Publishing, 1977.
36. Citizen's Nuclear Information Center, *Kenshō Tōdengenpatsu toraburugakushi* [*Inspection: Concealment of Trouble at Tokyo Electric's Nuclear Plants*], Iwanami Booklet no. 582, 2002.
37. Koide, H., *Genpatsu no uso [The Lies about Nuclear Power]*, Fusōsha, 2011, p. 119.
38. Berle, A.A. and Means, G.C., *The Modern Corporation and Private Property*, Macmillan, 1932.
39. Troncale, L.R., "Comparing Systems Pathology Treatments Across Systems Processes Theory", *Proceedings of the 55th Annual Meeting of the ISSS*, July 2011, p. 115.
40. Nuclear and Industrial Safety Agency, "Genshiryokuhatsudensho no hoan katsudō hyōka (shikō) no kekka ni tsuite" [Regarding (trial) evaluation of activities to maintain nuclear plants], June 2010.
41. Japan Nuclear Energy Safety Organization, "Jishin reberu 2PSA no kaiseki (BWR)" [Level 2PSA Earthquake Analysis], October 2010, pp. 4–7.
42. Reason, J., *The Human Contribution: Unsafe Acts, Accidents, and Heroic Recoveries*, Ashgate Publishing, 2008. *See also*, Reason, J., *Human Error*, New York: Cambridge University Press, 1990.
43. Chroust, G., "Regional Disasters and Systemic Reactions", *Proceedings of the 55th Annual Meeting of the ISSS 2011*, July 2011, p. 97.
44. Barnard, C.I., *The Functions of the Executive*, Harvard University Press, 1938, pp.44–45.
45. *Ibid.*, pp.55–57.
46. *Ibid.*, pp.261–270.
47. Simon, H.A., *Administrative Behavior: A Study of Decision-Making Processes in Administrative Organizations*, Macmillan Publishing, 1945, pp.80–82.
48. Barnard, C.I., *op. cit.*, pp. 302–305.
49. Drucker, P. F., *Innovation and Entrepreneurship*, Harper Collins Publishers, 1985.
50. Kawanaka, N., *Seisaku Kettei to Shakai Riron* [Policy Decision-making and Social Theory), Ryōsho Fukyūkai], 1984, pp. 16–19.
51. Swanson, G.E., *Social Change*, Scott Foreman and Company, 1971.
52. Swanson, G.A., "Management Observation and Communication Theory", *Proceedings of the 44th Annual Meeting of the ISSS*, July 2000, p. 121.

53. Maturana, H.R. and Varela, F.J., *Autopoiesis and Cognition: The Realization of the Living*, D. Reidel Publishing Company, 1980.
54. Polanyi, M., *Tacit Dimension*, Routledge & Kegan Paul, 1966. *See also*, Polanyi, M., *Personal Knowledge*, Routledge & Kegan Paul, 1958.
55. Maturana, H.R., "Ethics as Human Action—Ethics Belongs to the Emotional Domain", *Proceedings of the 44th Annual Meeting of the ISSS*, July 2000, p. 91.
56. Farrell, A., "Sustainability and Decision-Making", *Policy Studies Review*, Vol. 16: Nos.3/4 1999, pp. 37–43.
57. Arrow, K.J., *The Limits of Organization*, Norton & Company, 1974.
58. Cashore, B. and Vertinsky, I., "Policy Networks and Firm Behaviors", *Policy Sciences*, Vol.33: No.1., 2000, pp. 4–9.
59. Colborn, T., Dumanoski, D. and Myers, J.P., *Our Stolen Future,* The Spieler Agency, 1996.
60. UNAIDS, Global report: *UNAIDS Report on the Global AIDS Epidemic 2012*, 2012, pp. 6–8.
61. Bateson, G., *Steps to an Ecology of Mind*, Paladin, 1973.
62. *See*, Ministry of JUSTICE, A guide to the Corporate Manslaughter and Corporate Homicide Act 2007. *Available* at: http://www.hseni.gov.uk/guidetomanslaughterhomicide07.pdf#search='organizational+Manslaughter'. *Accessed* May, 2014.
63. Chroust, G., ICT support for disaster management. In Doucek, P., Chroust, G. and Oskrdal, V. (editors), *IDIMT 2012 ICT-Support for Complex Systems,* Trauner Verlag Linz, *vol.38* Sept 2012, pp. 13–23.
64. Watsuji, T., *Ningen no Gaku toshite no Rinrigaku* [Human Learning as Ethics], Iwanami Shoten, 1934, pp. 46–63.
65. Sandel, M.J., *What Money Can't Buy: The Moral Limits of Markets*, Farrar, Straus and Giroux, 2012. *See also,* Sandel, M.J., *Justice: What's the Right Thing to do*, Farrar, Straus and Giroux, 2009.
66. Bertalanffy, L.V., *General System Theory*, George Braziller, 1968, pp.47–48.
67. Krauss, L.M., *A Universe from Nothing: Why There is Something Rather than Nothing*, Simon & Schuster Ltd, 2012.
68. Barnard, C.I., *op. cit.*, pp.46–48.
69. Wiener, N., *Cybernetics; or Control and Communication in the Animal and the Machine*, MIT Press, 1948.
70. Shiozawa, Y., *Fukuzatsu-kei Keizaigaku Nyūmon* (Economies of Complex Systems), Seisansei Shuppan, 1997.
71. Whitehead, A.N., *Process and Reality: Science and the Modern World*, Macmillan, 1926.
72. Murata, H., *Kanri no Tetsugaku —Zentai to Ko, Sono Houhou to Imi* [Management Philosophy—the Collective and the Individual, Methods and Meaning]. *See also*, Yamamoto, Y., *Keieigaku Honshitsuron* [Essential Theory of Management Science], Moriyama Shoten, 1961.
73. Arrow, K.J., *op. cit.*
74. Drucker, P.F., *The Essential Drucker: The Best of Sixty Years of Peter Drucker's Essential Writings on Management*, Harper, 2001.
75. Tsukio, Y., *Journeys beyond Time and Space: The Wisdom of the Indigenous Peoples*, Asahi eco books, 2012.
76. Diamond, J.M., *The World Until Yesterday: What Can We Learn from Traditional Societies?*, Viking Penguin, 2012.
77. Atsuji, S., *Soshiki Kettei no Kagaku* [Science of Organizational Decision-making], Kansai University Press, 1999. *See*, Polanyi, M., *The Tacit Dimension,* Routledge & Kegan Paul, 1966. *See also*, Ayer, A.J., *The Foundations of Empirical Knowledge*, Macmillan, 1940. *See for example*, Niwamoto, Y., "Kindai kagakuron wo koete", *Osaka Shougyou Daigaku Ronshu* No.66, 1983, pp.119–131.
78. Gazzaniga, M.S., *The Social Brain*, Basic Books, 1985. *See also*, Russell, P., *The Global Brain*, Tarcher, 1983.
79. Kineman, J.J., "Crisis sciences for sustainability", *Program & Abstracts 2015,* (ISSS Journals), pp.166–167.

80. Piketty, T., *Capital in the Twenty-First Century*, Harvard University Press, 2014. (*Le Capital au XXIe siècle*)
81. Diamond, J.M., *The World Until Yesterday: What Can We Learn from Traditional Societies?*, Viking Penguin, 2012. *See also*, Lovins, A.B. et al., *Saiseikanō Energy ga Hiraku Mirai [A Future Built on Renewable Energy]*, Iwanami, 2013.
82. McNeill, W.H., *A World History*, Oxford University Press, 1998.
83. Lloyd, C., *What on Earth Happened?: The Complete Story of the Planet, Life and People from the Big Bang to the Present Day*, Bloomsbury Publishing Plc, 2012.

Chapter 9
Remaking Eco-civilization by Sustainable Decision-Making

9.1 Eco-citizenship Beyond the Generations

Is there something in humanity's nature that makes us seek out unsafety and risk? Already in the hunter-gatherer society of prehistory, a pattern of communal living for the sake of survival and cooperation to obtain food, clothing, and shelter was universal. Well before recorded history, as humankind progressed from the New Stone Age to the Bronze Age and then the Iron Age, groups of people specialized in making farming tools, weapons, horse tackle, and other forms of metalworking and needed a system of cooperation at the level of the village, community, and state. Meanwhile, from the time of the first agricultural revolution, ruling the waters, for instance by predicting the timing of river floods to assist with seed sowing, tilling, and harvesting, was the basis for the existence of the state.

In Ancient Egypt it was the River Nile, in Mesopotamia the Tigris and the Euphrates, in the Indus Valley the Ganges, and in China the Yellow and Yangtze Rivers, but control of the waters to avoid the natural disasters of flooding and drought was the essence of control of the land. To take control of the waters was to guarantee the harvest of agricultural produce, and the ability to forecast natural disaster was the origin of state power. Conversely, in ancient civilizations, had there been no natural disasters, it is uncertain whether the state could have acquired power. In agricultural societies, to maintain the state monopoly of power, a fiscal basis was established through collection of tributes and taxes. It became so that to rule the waters was to rule the land, and the response to natural disaster undeniably helped promote the birth of the state.

Thanks to the progression of farming tools from wood to stone and the invention of bronze farming tools to replace iron, wheat, maize, and rice yields surged in the 'agricultural revolution' beyond the hunter-gatherer society. From the beginning of historical times—along with the technological innovation of metalworking—stone, papyrus, bamboo, and wooden tablets were used to record taxation (taxes and tributes), and the social division of labor progressed. With the move from bronze

S. Atsuji, *Unsafety*, Translational Systems Sciences 7,
DOI 10.1007/978-4-431-55924-5_9

to iron, the quality of arms and weapons improved and the capacity and opportunities for combat were extended. Meanwhile, ironworking brought the shipbuilding technology that made possible long-distance seafaring and the manufacture of artillery, presaging the advent of 'the age of exploration'. Continuing up to the industrial revolution, the state monopoly of power made use of iron and woodworking technology to plunder overseas wealth. The invention of the internal combustion engine based on steam heat and fossil fuel (coal and petroleum) wrought dramatic change in the subsequent 'industrial revolution'.

Trains and ships driven by the steam engine and other vehicles running on petroleum and internal combustion engines, coinciding with T. Edison's 'electric revolution' and fusing with transportation technology, fed into the 'management revolution' that started with H. Ford and F. W. Taylor. The internal combustion engine and electric lighting and motors enabled the change from home-based workshop manufacture to factory-based mass production, while the securitization of funds through the stock-company system made possible mass production. As a result, the concept of the 'going concern' gave birth to the new social organization of the commercial enterprise. Having brought about this fusion of funds and technology, modern capitalism, building on the foundation of the stock-company system with stock-option exchange securities and supplied through the medium of the financial markets, led the development from the market economy to industrially organized society and onward toward the market society.

As shown in the world scholarly locus (Table 9.1), following the American Civil War and under the influence of F. W. Taylor's 'scientific management', the next trend of human relations involved in Western Electric's Hawthorne experiment (E. Mayo and F. Roethlisberger) focused strongly on the relationship between the commercial enterprise and the worker from the viewpoint of human relations, as part of the 'management revolution'. After the great world depression of 1929, the relation of the organization to the individual was reviewed from the viewpoint of effectiveness and efficiency, crystallizing in the grand theory of the modern organization developed by C.I. Barnard, M.P. Follet, and H.A. Simon.

The industrial revolution was succeeded by the management revolution. Amid the favorable economic climate of the times, mass production revealed a dark side of unsought consequences in the form of mine explosions, factory accidents, and other unnatural disasters, which came to be seen as an issue. In Japan, many workers and local residents became victims of pollution at Chisso Minamata, Sumitomo's Besshi copper mine and the Ashio copper mine. In the initial stage of modern capitalism, there was an endless stream of accidents and disasters, and the combination of Japanese 'industrial revolution' and 'management revolution' not only drove a wider gap between rich and poor through greater labor efficiency, but wealth imbalance also became the cause of labor disputes, demonstrations, and riots. The gap opened up by the technological power and organizational power of capital increased the economic power of the commercial enterprise and the financial power of the state. On the other hand, the social discrimination caused by the expansion of economic inequality also became a breeding ground for unsafety and today continues to feed the global north–south wealth issue.

Table 9.1 World scholarly locus

Sustainalism		T. Piketty (1971–)	J.M. Diamond (1937–)	L.R. Brown (1934–)
Eco-civilization Revolution		G. Chroust (1941–)	H.R. Maturana (1928–)	M.J. Sandel (1953–)
		J. Reason (1938–)	T. Colborn (1927–)	A.B. Lovins (1947–)
Bio-Revolution	S. Yamanaka (1962–)		L.M. Silver (1952–)	E.J. Chaisson (1946–)
Energy Revolution		L. Watson (1939–2008)	E.A. Feigenbaum (1936–)	K.E. Weick (1936–)
	H. Itsuki (1932–)	M. Foucault (1926–1984)	M.L. Olson (1932–1998)	J. Naisbitt (1929–)
ICT Revolution	H. Yukawa (1907–1981)	I. Prigogine (1917–2003)	A. Toffler (1928–)	E.H. Schein (1928–)
		W. H. McNeill (1917–)	J.G. March (1928–)	P. Krugman (1953–)
		A.M. Turing (1912–1954)	D.H. Meadows (1941–2001)	S. Beer (1926–2002)
		J.L. Monod (1910–1976)	T. Kuhn (1922–1996)	G.E. Swanson (1922–1995)
		C. Lévi-Strauss (1908–2009)	H. Kahn (1922–1983)	K.J. Arrow (1921–)
		M. Merleau-Ponty (1908–1961)	R.M. Cyert (1921–1998)	D. Bell (1919–2011)
Scientific Revolution		W. Leontief (1905–1999)	J.G. Miller (1916–2002)	H.A. Simon (1916–2001)
	T. Watsuji (1889–1960)	W.R. Ashby (1903–1972)	C.E. Shannon (1916–2001)	P.A. Samuelson (1915–2009)
	K. Nishida (1870–1945)	L. von Bertalanffy (1901–1972)	M. Friedman (1912–2006)	P.F. Drucker (1909–2005)
		E. Fromm (1900–1980)	J.K. Galbraith (1908–2006)	A.H. Maslow (1908–1970)
		N. Luhmann (1927–1998)	R.L. Carson (1907–1964)	G. Bateson (1904–1980)
Management Revolution		M. Polanyi (1891–1976)	J. von Neumann (1903–1957)	T. Parsons (1902–1979)
		H.W. Heinrich (1886–1962)	S.K. Langer (1895–1985)	A.A. Berle, Jr. (1895–1971)
Social Revolution		J.A. Schumpeter (1883–1950)	C.I. Barnard (1886–1961)	N. Wiener (1894–1964)
		B. Russell (1872–1970)	A. Einstein (1879–1955)	E. Mayo (1880–1949)
		M. Weber (1864–1920)	A.N. Whitehead (1861–1947)	M.P. Follett (1868–1933)
Industrial Revolution			T.A. Edison (1847–1931)	F.W. Taylor (1856–1915)
		K. Marx (1818–1884)		
		G.W.F. Hegel (1770–1831)		
Colonial Period		I. Kant (1724–1804)		
Age of Discovery		I. Newton (1642–1727)		
	M. Miyamoto (1584–1645)	Galileo Galilei (1564–1642)		
	Wang Yangming (1472–1529)	N. Copernicus (1473–1543)		
		Leonardo da Vinci (1452–1519)		
Epoch	Mencius (372BC–289BC)	Aristotle (384BC–322BC)		
		Plato (427BC–347BC)		
		Socrates (469BC–399BC)		
	Confucius (551BC–479BC)	Pythagoras (582BC–496BC)		
	Asia/Oceania	Europe/Africa	America	

Modern capitalism, in concert with the stock-company system, enabled the procurement of funds from the financial market, creating a social structure centered on the market economy. On the other hand, the damage which the money-worshipping market society has inflicted on the individual's sense of values and on social conventions is now coming to light. Amid the current economic materialism, we have lost the spiritual and psychological fulfillment which depended on the safety and security that have been sacrificed. The social functions that support contemporary society—organizational-management functions and public-policy functions—have fostered collusive relationships between political power and the economically wealthy in the market society, and, due to the structural inertia of organizations, they are now drifting in a direction opposite to public welfare. But nowadays, happiness and satisfaction are more sought after than material goods. A light has now been shone into the spiritual darkness resulting from the loss of balance, and a debate has begun from the standpoint of a reaction against materialism.

Following the material wealth brought to the developed countries by the industrial revolution, a return to spiritual wealth has begun. Already, at the time of the 'scientific revolution', there were scholars pointing to the unfeeling nature of the logic of an organized society centered on industry and the economy. In the age of the scientific revolution, these were mainly T. Kuhn in *The Structure of Scientific Revolutions*, J. Monod in *Le Hasard et La Nécessité*, W.R. Ashby in *The Design for a Brain*, and N. Wiener in *Cybernetics*, as well as D. Bohm and G. Bateson, while concern over the effect of science and technology on human society was evident even in A. Einstein's *Theory of Relativity*, B. Russell's *Principles of Mathematics*, and in A. Turing, J. von Neumann, and C. E. Shannon. This recourse to humanity provoked by the scientific revolution can be seen as representing a renaissance focused on the relationships between science, humankind, and society. Even in this scientific revolution, however, human-made disasters and accidents were seen as irregular phenomena and not as the objects of scientific inquiry. Not only unnatural disasters but even natural disasters were insufficiently recorded, so that wide-ranging instances of unsafety have historically been disregarded by people and the state as negative phenomena, allowing the same human-made disasters and accidents to be endemically repeated.

In *Guns, Germs and Steel*, J. Diamond says that humanity stands at a parting of the ways where it must choose either the horse of destruction or the horse of sustainability. Looking back over the history of humankind, we find that China's Spring and Autumn Period, the Mayan civilization, and the ancient Egyptian and Mesopotamian civilizations all used an astronomically calculated calendar to predict floods and droughts and warn farmers of natural disaster, thus laying the foundations of the state.

As capitalism is rumored to be failing, the Christian Protestantism that swept the world in the modern age is today in increasingly serious confrontation with the Islamic extremism. Modern capitalism has been introduced to Russia, China, Vietnam and other socialist and communist countries, expanding by fitting in with collectivism. In the history of management scholarship, the scientific theories of F. W. Taylor, E. Mayo, C. I. Barnard, H. A. Simon, P. F. Drucker and other representatives of strategic domains of American management science who contributed to shaping modern capitalism have gradually been pushed aside by the changing times and fallen into disregard. As an alternative new line of thought, L. Brown, and others including J. Diamond, T. Colborn, L. Silver, M. Sandel, and J. Reason have created a new fact-based academic current through empirical research based on presentation of hypotheses that can be understood by people worldwide. Management theories that have contributed to the development of the modern organization have successively shifted their focus from the conventional 'going concern' to the 'public organ' and thence to 'sustainability'. Similarly, in the modern age, led by stakeholders with widely different interests, there has been an evolution from the profit-seeking academic doctrines of the money-worshipping individual organization to a situation where the social functions of 'compliance and governance' have come to be required of organizational management. The academic orientation too, notably in A.A. Berle, has shown a paradigm shift toward pursuit of benefit in terms of social well-being.

In conventional economic and management research, the mainstream approach has been directed at securing individual and national wealth based on growth theory, but in the modern age, managerial strategies in the global market have begun to take account of impacts on the social environment and the ecosphere. Among the particular examples referred to here are the alarm sounded by L. Brown over the global environment, J. Diamond's civilizational collapse, T. Colborn's environmental hormones (brain contamination through accumulation of chemical substances), the dangers posed from bio-business to moral hazard as described by L. Silver, and M. Sandel's concern over the moral limits (moral dilemmas) accompanying the shift from a market economy to a market society. These are all part of an ongoing paradigm shift from growth theory to an emphasis on survival. As once foreseen by C. I. Barnard in his reference to 'unsought consequences' and P. F. Drucker in his 'unexpected results', the sum total of organizational activity based on human cooperation is now being examined in terms of the actual benefit for human society which results from it (Table 9.2).

The ideas and insights of these earlier scientists are carried on through the ISSS in works such as G. Chroust's 'human-made disaster', L. Troncale's 'systems pathology', and J. Kineman's 'crisis sciences', as well as J. Reason's 'organizational accident' and S. Bennett's 'disaster management', which show a move toward interdisciplinary research oriented to the resolution of problems associated with modern 'unsafety phenomena'. These are logical progressions of the thought of Barnard and Drucker in the historical development of management science and indeed belong to a trend toward global environment-oriented eco-management centered on a shift from growth to 'sustainability'. In recent years, crisis management and sustainability has been identified as a focus issue within IFSAM, a conspicuous strand of which is the academic current toward human resource management (HRM) and crisis/risk management from the perspective of global resources for the sake of social and eco-survival as advocated by M. Morley and others, the new trend of management scholars.

In the Fukushima nuclear disaster (Chap. 2), the response by the Japanese government and the power company revealed an approach led not by the principle

Table 9.2 Genealogy of management theories

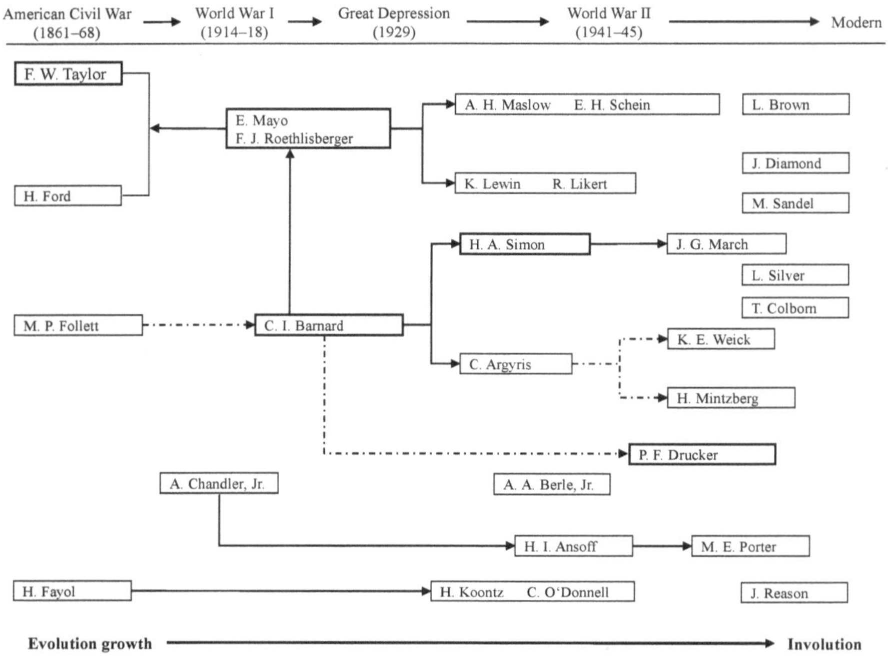

Note: Drawing based on 'Scholarly history of management' by H. Koontz, C. O'Donnell, and H. Weihrich

of protecting the public, but instead by the priority given to self-protection by government officials and the management teams of the commercial enterprises. It is undeniable that, where its interests coincided with those of the power company, the government, rather than protecting the public from radioactive contamination, sacrificed the public by working to save face and maintain the power structure. Contemporary society is thus upside down. The original irrational decision to locate a nuclear power station in an earthquake-prone area, where it was allowed to age for more than 40 years, was compounded by the obstinate cover-up of successive operational failures and accidents and the falsification of records, which have since been revealed and called into question. It cannot be ruled out that the egregious irrationality and obstinacy surrounding the Fukushima nuclear disaster will be found to have damaged the mental health of the children of the next generation. As for the next generation, who will live in this country in the hope of social justice—what should they believe? In a 2014 survey, the residents of the earthquake zones were found to suffer five times the normal rate of depressive symptoms, indicating a spreading spiritual darkness.

In Germany and Italy, in the wake of the Fukushima nuclear disaster, policy decisions have already been taken to move toward denuclearization, with the decommissioning of nuclear reactors in the near future. In Japan itself, however, nuclear power issues have been shielded within an archaic social structure which upholds the interests of the nuclear industry and its regulatory structure and the associated approval and licensing system and protects them within a Galapagos-like cultural ethos. In public opinion research, 70–80 % of the Japanese public supported the denuclearization of energy. However, the underlying approaches of "Japanese spirit with Western learning", which divides science and technology from spiritual culture, may have obscured the nuclear power issue. The social functions of 'organizational management'—on one hand, that is the strategic decision-making designed to control the technology and resources of the private sector, and on the other hand 'public policy' systems to supervise private sectors—are in a process of divergence from the will of the public, like that of the JR West Accident (Chap. 4). The politicians who are in charge of public policy and the National Diet members who represent influential lobbies come to curry favor with voters only at election time. The very next day, they fall back in step with the operators of the commercial enterprises that assist with their election funds, representing only the interests of the financial world, and committing a considerable amount of misappropriation. Some commentators claim that even the media are in the pocket of vested interests. Surely I am not the only one who feels that Japan, the country which experienced the Fukushima catastrophe, is now in urgent need of 'citizenship power'?

Who could have imagined that the chlorofluorocarbon gas used in refrigerator coolant would open a hole in the ozone layer of the global stratosphere, allowing cosmic radiation to reach the earth's surface and presenting the threat of skin cancer and other issues, especially in the southern hemisphere? No one foresaw this decisive damage to the global ecosphere (Gaia). It is the same story with environmental hormones (Chap. 3) that accumulate and cause damage to children's brains, and with the socio-biological hazard (Chap. 5) of the viruses encouraged by global warming that are extending their range toward the north and south poles. Nor was it foreseen that humans would be infected with measles, tuberculosis, and smallpox from cattle, whooping cough from pigs, influenza and malaria from poultry, or any of the other infectious diseases passed on from livestock. Global warming (Chap. 6) itself is the accumulation since the industrial revolution of CO_2, smoke and soot, and PM2.5 from fossil fuels coal and petroleum, which, according to WHO survey results, is related to the high dioxin levels detected in breast milk in Europe. Moreover, the cumulative effects of CO_2-based atmosphere global warming, and hydrosphere ocean warming by thermal effluent from the cooling process of the world's 441 nuclear reactors, which cause the melting of the frozen soil of Greenland, Siberia, the Arctic, and Antarctic, and the release of the methane hydrate fixed in the continental shelf seabed, produce at least 20 times the greenhouse-gas effect of CO_2, further exacerbating global warming. This development could combine with food crises in the wake of crop failures due to climate change and abnormal weather patterns, which would be further worsened by the population explosion (Richard Muller, Berkeley Earth Surface Temperature Project).

Originally, the role of the state was to protect its populations by preventing disaster- and accident-related risk and crisis, ensure fiscal capacity through taxes collected, and work to prevent disaster in the public interest. The state itself did not engage in direct creation of wealth and services and did not produce commercial and industrial profit. If, as in the Fukushima nuclear disaster, risk and crisis are hidden from the public, the loss of the state's power base naturally follows, which should have been clear from the fall of the Berlin Wall. A similar trajectory was reflected in the following Japanese general election with the huge defeat of the Democratic Party of Japan and the recovery of the Liberal Democratic Party. Under the state apparatus of modern capitalism, a collusive structure is appearing between the two great social functions of public policy to formulate political strategy and business administration to operate commercial enterprises. These social roles are normally in opposition and mutually restraining, but in the modern situation where the social function of disaster prevention for public protection is weakened, how can this embrace be unraveled? For the business administration and public policy on which we depend in the modern age to operate as social functions, it is surely essential to return to the starting point and rediscover 'citizen power'. Many international intellectuals concerned with the public interest, notably L. R. Brown and J. M. Diamond, have made clear statements on their hopes for the potential of 'eco-citizenship'.

The so-called revolution in information and communications technology (ICT), which fuses the digital processing and telecommunication that are features of the Internet society, has created electronic space and encouraged creativity, but on the other hand has also given rise to 'cyber terror' and other threats to information resources and privacy. There are companies who make a business of the commodification of privacy: that is, they entice readers to consent to online terms and conditions intentionally made difficult to read, then have them press a 'like button' on Facebook to automatically broadcast their date of birth complete with photograph to an indefinitely large audience. Still other companies such as Google have the aim of 'capitalizing knowledge': there are Wikipedia and so on, who digitalize the data of physical libraries and apply charges to what were charge-free information-search sites. Digitalization in the Internet society produces globalization and impacts the world economy, politics, and public order. At the same time, whether for good or for ill, conflict and terrorism are increasingly internationalized. The issues of nuclear disaster and energy which were presented to the world by Fukushima have revealed the possibility of the globalization even of human-made disasters.

Turning our attention to the energy resources of the current age, conflict over such resources, which has been the cause of world war and conflict in the past, appears to be being mitigated by the renewable energy revolution. Renewable energy sources such as solar and wind power, geothermal energy, biomass, hydrogen fuel and sources as diverse as oil sand, shale gas, methane hydrate, and artificial photosynthesis are being explored. In some cases, for instance, in the US city of Portland and at Feldheim in Germany, an eco-energy shift is being attempted. Meanwhile, in a separate trend arising in the developed countries with their aging populations, regenerative medicine (ES, iPS, and STAP phenomena), genetic diagnosis, and other developments signal the start of a 'bio-revolution' affecting health and longevity issues. As pointed out by L. Silber, the organ business, Nobel Prize winner sperm banks, surrogate motherhood and other 'bio-businesses' are emerging. In tandem, there are voices raising concerns over the ethics of bio-engineering.

Based on the logic of capital, the bio-revolution occurring simultaneously with the 'energy revolution' contains implicit issues of bioethics and social convention. For humankind, the 'holy trinity' of the energy, bio-, and ICT information revolutions was originally supposed to serve the public interest, but despite the expectation of a benefit from their progression, the resulting gap between communities and individuals introduces a latent inequality. The current world population is exploding toward a figure of ten billion (2050) estimated by the United Nations. Food crises, water shortages, and conflict over energy resources and other issues are spreading across the globe, and the water, air, food, and energy that humanity needs to live are being stretched beyond their limits. The depletion of all resources on a scale never yet experienced in human history, with a few exceptions such as the Easter Island and Mayan civilizations, is exceeding what we envisaged. But this unsafety has been forgotten in the generational transition as we have sought escape by turning away from the alarming phenomena of crisis and risk, which means that they may well return like a boomerang in the form of a fresh major misfortune.

Going forward, the trend toward 'energy revolution' and 'bio-revolution' arising from the 'ICT revolution' will bring not only economic profit and social surplus but also a wealth gap and simultaneously not only the globalization of the effluent and exhaust gases emitted by big capital's mega-factories and the associated by-products and goods and services, but also the globalization of disasters and accidents. P. F. Drucker's 'unexpected result' presents a besetting problem, especially for CO_2-emitting countries such as China and the United States. The logic of capital under modern capitalism has spread worldwide across national boundaries, but people are beginning to realize that, however hard they pursue profit, it does not make them particularly happy. Through a system focused on commercial profit in line with the logic of capital, the multinational and world enterprises supported by big capital have brought about a conversion of the market economy into a market and world society, which has awakened in people an 'awareness of sustainism' in preference to capitalism. The basic principle of organizational management must become to maintain the character of the enterprise as a going concern and a public organ, and to realize sustainability. In today's world, compliance, governance, and human resource management designed to serve social well-being are essential. In public-policy formulation too, the systems for medical treatment, welfare, health, pensions, and nursing care, which are a common difficulty faced by developed countries, have become dysfunctional, and their maintenance has become challenging in countries that continue to operate a national deficit. What will we reap in the near future from the holy trinity of ICT revolution, energy revolution, and bio-revolution?

As foreseen by L. Silver, the incongruence of the effects of bio-revolution exemplified in bio-business and by T. Colborn and brain damage from environmental hormones illustrates the undeniable possibility that human-made disasters may mutate by accumulation into natural disasters. We can probably infer that going forward the world will continue to be oppressed by the collusive relationship between private administration and public policy, that the gaps between north and south and east and west will grow, and that instability will increase progressively. The imbalance arising from the inequality of wealth and material possessions is the very source of unsafety, and cannot be remedied under present conditions. To arrest this development, we can look to citizen power in opposition to the public policy and business management of the current age, which foreshadows a new global revolution in favor of an eco-civilization.

There are societies where those who advocate education for girls are shot, like the schoolgirl Malala in Pakistan, and there are children who take up a gun after losing their families to war and conflict or terror. Meanwhile, there are children receiving education in economically rich societies and leading quite different lives. True to the warning that all empires must come to an end, this kind of imbalance will before long come back to haunt the developed countries. Unsafety arising from imbalance brings human-made disasters and combines with natural disaster to precipitate catastrophes. These can visit the people of both developing and developed countries, who share the same planet, and to whom the saying "there but for good fortune may go you or I" applies universally. From Europe's floods and America's snow blizzards to Africa's drought and Asia's extreme heat, "what happened next door yesterday" will be visited on us today under an altered guise. Like natural disaster, human-made disasters happen to everyone. History teaches that things happening in distant locations will sooner or later have repercussions at home. Sources of great suffering must therefore not be treated as other people's problems but must be prevented by humans working together at the global level. To prevent the great misery of civilizational collapse, we must take our chance on a 'global eco-civilization' based on eco-citizenship in opposition to business management and public policy, by shifting from 'capitalism to sustainism'. Since we cannot ask this of the people of the future, nor space aliens of course, we, as members of Gaia, must in the end overcome these issues ourselves.

The scope of systems sciences based on interdisciplinary translational research focuses on unsafety as it transmutes from the physical and the biological to civilizational collapse.

9.2 Translational Research for Climate Crisis

Human history is the history of disaster and is intimately entwined with the history of the earth and the history of the universe. As is clear from the chronological events, the period of global warming within the hyper-timescale of the Earth has seen frequent human-made disasters such as wars and conflicts. At other times, the solar cycle or dust and ash clouds from meteor strikes or volcanic eruptions have blocked the sun's rays and triggered cooling, leading to sustained crop failures and nomadic migrations caused by loss of pasturage, for instance the southward migration of the Germanic tribes and the westward movement of the Huns. The result has been hunger and famine leading to uprisings and coups d'état. Events in the history of the earth and human history thus have a strong degree of interconnection, such that natural disaster may lead to unnatural disaster (Table 9.3).

Table 9.3 Translational review between earth and human history

BC5500–AD1300

Year	History of the Earth	warm/cold	History of Humankind
5500BC	Sahara desertification	global cooling	
5000BC		global cooling	Yellow River civilization
3500BC		global cooling	Mesopotamian civilization
3000BC			Egyptian civilization
2600BC		global cooling	Indus Valley civilization
2300BC	Global cooling (–2100BC)	global cooling	Fall of Egypt's Old Kingdom
1800BC			Decline of Indus Valley civilization
1760BC			Code of Hammurabi
1750BC	Veniaminof volcano		
1627BC	Santorini eruption		
1600BC		global cooling	Establishment of Shang dynasty
1500BC	Global cooling (–1000BC), great worldwide drought	global cooling	
1200BC		global cooling	Olmec civilization
1159BC	Mount Hekla eruption	global cooling	
770BC			Spring and Autumn Period/Warring States Period (–221BC)
627BC			Assyrian Empire
428BC			Buddhism
331BC			Eastern expedition of Alexander the Great
221BC		global warming	Unification of China under Qin state
264BC		global warming	Punic Wars
217BC	Vesuvius eruption	global warming	
27BC		global warming	Establishment of Roman empire
30		global warming	Christianity
220		global cooling	Three Kingdoms Period (–280)
250		global cooling	Maya civilization
300		global cooling	Northern tribes invade northern China
		global cooling	Spread of Hinduism (–400)
365	Earthquake on Crete	global cooling	
370		global cooling	Western migration of Huns
375		global cooling	Great southern migration of Germanic tribes
476			Fall of Western Roman Empire
526	Earthquake at Antioch, Turkey	global cooling	
610		global warming	Islam
750			Establishment of Abbasid Caliphate
962		global warming	Establishment of Holy Roman Empire
1096		global warming	First Crusade
1206		global warming	Establishment of Mongol Empire
1257	Samalas eruption	global warming	
1299		global warming	Establishment of Ottoman empire
		global warming	Flourishing of Aztec and Inca empires

■ global warming ■ global cooling

(continued)

Table 9.3 (continued)

1301–1800

Year	History of the Earth	warm/cold	History of Humankind
1315–17	Great famine in Europe		
1334–37	Famine in China		
1337			Hundred Years War between England and France
1346–50	Plague outbreak in Europe (Black Death)		World population near 370 million
1452	Kuwae eruption		
1509	Istanbul earthquake		
1517			Reformation
1521			Fall of Aztec Empire
1533			Fall of Inca Empire
1556	Huaxian earthquake		
1571			Battle of Lepanto
1581			Netherlands independence
1586	Mount Kelud eruption		
1600	Huaynaputina eruption		
1641	Mount Parker eruption		
1642			English Civil War
1644			Demise of Ming dynasty, China united under Qing dynasty
1668	Anatolian earthquake		
1669	Mount Etna eruption		
1672	Mount Merapi eruption		
1688			Second Hundred Years War between France and England
1707	Hōei earthquake, Mount Fuji eruption		
1711	Mount Awu eruption		
1737	Kolkata earthquake and cyclone		
1746	Lima earthquake and tidal wave		
1755	Lisbon earthquake		
1757			Battle of Plassey
1770	Bengal famine		
1775			American War of Independence
1776			American Declaration of Independence
1780	Lesser Antilles Hurricane		
1783	Mount Laki eruption		
1789			French Revolution
1792	Mount Unzen eruption		

(The warm/cold column is shaded global cooling throughout 1315–1792.)

■ global warming ■ global cooling

(continued)

Table 9.3 (continued)

1801–1930

Year	History of the Earth	warm/cold	History of Humankind
1804		global cooling	Haitian Revolution
1815	Mount Tambora eruption	global cooling	
1819		global cooling	Britain acquires Singapore
1831		global cooling	First Turco-Egyptian War (–33)
1839		global cooling	Ottoman Empire begins *Tanzimat*
1840			Opium War (–42)
1845	Famine in Europe		
1846			Mexican-American War (–46)
1849			British conquest of India
1851			Taiping Rebellion (–64)
1853			Crimean War (–56)
1854	Ansei earthquakes		
1856	Mount Awu eruption		Arrow War (–60)
1857			Indian Rebellion (–59)
1861			American Civil War (–65)
1868	Chile earthquake and tidal wave		
1869			Completion of Suez Canal
1877			Russo-Turkish War (–78)
1877	Drought in China (–78)		Establishment of British Indian Empire
1879	Drought and famine in India (–89)		
1883	Krakatoa eruption and tidal wave		
1887	Yellow River floods		
1891	Nōbi earthquake		
1894			Sino-Japanese War (–95)
1896	Sanriku earthquake		
1898			Spanish-American War
1899			Boer War (–1902)
1900	Drought and famine in India		
1902	Mount Pelée eruption	global warming	
1904		global warming	Russo-Japanese War (–05)
1908	Messina earthquake	global warming	Young Turk Revolution
1911		global warming	Xinhai Revolution
1914		global warming	**First World War (–18)**
1917		global warming	Russian Revolution
1919		global warming	Treaty of Versailles
1922	Chile earthquake and tidal wave	global warming	
1923	Great Kantō Earthquake	global warming	
1929		global warming	Great Depression

■ global warming ■ global cooling

(continued)

Table 9.3 (continued)

1931–2014

Year	History of the Earth	warm/cold	History of Humankind
1932	Caucasus region drought and famine (–34)		
1933	Sanriku earthquake and tidal wave		
1937			Sino-Japanese War (–45)
1939			**Second World War (–45)**
1944	Tōnankai earthquake and tidal wave		
1946	Aleutian Islands earthquake		
1947			Independence of India and Pakistan
1950			Korean War, World population over 2.5 billion
1955			Warsaw Pact agreement
1955			Asian-African conference
1960	Great Chilean earthquake		Independence of African nations
1960			Vietnam War (–75)
1962			Cuba crisis
1964	Alaska earthquake		
1970	Peru earthquake		
1973			First oil shock
1979			Iranian Revolution, Second oil shock
			Soviet invasion of Afghanistan (–89)
			Three Mile Island nuclear power disaster
1980			Iran-Iraq War (–88)
1982	El Chichón eruption		
1986			Chernobyl nuclear power disaster
1989			Tiananmen Square incident
1991	Mount Unzen eruption, Mount Pinatubo eruption		Gulf War, dissolution of Soviet Union
1995	Great Hanshin earthquake		Population explosion
2001	Global warming		America 9/11 terror attacks, Terrorism
2003			Afghan War and Iraq War
2004	Sumatra earthquake		Global Terrorism
2005	Hurricane Katrina		JR Fukuchiyama Line Derailment Accident
2008	Sichuan earthquake		Bankruptcy of Lehman Brothers, Financial crisis
2010	Chile earthquake and tidal wave		Arab Spring
	Eruptions of Eyjafjallajökull		Conflicts Expantion of North Africa
2011	Great East Japan Earthquake		Fukushima nuclear power accident
2012	Abnormal weather		Syrian Civil War
2013	Super Typhoon Haiyan, Chelyabinsk meteor		World population over 7.2 billion
2014	Mt. Ontake eruption		Ukrainian revolution
2015	Floods and drought in the world		Paris terror attacks, EU refugee flow
			Climate change, food and water crisis
2050			World population over 10 billion (estimate by UN)

■ global warming ■ global cooling

Note: Based on Super Time Scale of the Earth by Japan Society of Geology and a World History by W. H. McNeill.

In the environment, not only the water sphere (seas, rivers, lakes, and water vapor) but also the atmosphere, the land, and the earth's crust are fundamentally linked in a state of continuous circulation. Additionally, cosmic radiation from the galaxy and from supernova explosions in outer space, solar activity cycles, and meteor strikes create clouds in the earth's atmosphere that cause climate change (*see also,* connecting Chap. 1 Table 1.1, above). These changes in the history of space have not only had a determining influence on global cooling and global warming but were also connected with the cyclical mass extinctions in geological time, for instance that of the giant insects (50–100 times larger than now) and the large reptilians (5–10 times bigger). The relationship between the history of space and the history of the earth has been decisive in human history, from the Ice Age, through prehistory and the age of civilization to the modern age. In this context, the unnatural disasters which we encounter in our normal life spans are events of a single moment's duration. For people living today, disasters may bring great misery and appear fateful, but in the light of human wisdom we should attempt to see them in proportion. Just as natural disaster may underlie human-made disaster, equally the reverse effect may apply. It is the aim of this book to point out that the totality of the activities of human civilization can in this way trigger human-made disaster that leads on to natural disaster.

Recent years have seen global warming and abnormal weather patterns. Hurricanes, great tornadoes, and extreme cold have struck America while Europe has been hit by low pressure systems causing violent storms and major flooding alongside avalanches and heat waves. In addition to major earthquakes in Haiti, New Zealand, Sumatra, Turkey, Japan, and China, earthquakes have also been recorded in France, while volcanic eruptions have occurred in Iceland and Japan, great floods in Thailand and Bangladesh, and a super-typhoon in the Philippines. Drought and water shortage plague Africa, while acid rain in South America and Europe brings deforestation and desertification. On top of these comes the accelerating depletion of water resources worldwide, leading to the spread of hunger and famine. The list is endless. As well as such natural disasters, global warming has encouraged the poleward expansion of infectious and endemic diseases, while the movement of people promotes the worldwide expansion of infectious tropical diseases such as Ebola hemorrhagic fever, West Nile fever, Dengue fever, SARS, and MERS. Meanwhile, the wealth gap between the northern and southern hemispheres and religious tensions between Christianity and Islam have opened up multiple sources of terrorism and ethnic conflict. Past human history actually indicates a pattern of globalized human-made disaster through war and conflict in periods of warming. In periods of global cooling, such as those caused by meteor strike, volcanic eruption, or supernova explosion, after which the globe becomes a 'snowball', life activity has been suppressed, and new environments have appeared to which selectively adapted living species have evolved. There are also phenomena running counter to this natural process.

The great calamities that the author has experienced, the Great East Japan Earthquake and the Great Hanshin Earthquake, the eruption of Mount Ontake, the Fukushima Nuclear Power Station disaster and the JR West rail accident, show a connecting cyclicality and commonality. Of course, natural disasters and unnatural disasters are phenomena which differ in essence and in logical character, but in terms of experience and statistics, the two types of disaster frequently coincide and can be seen in the chronological table to exhibit connectedness and cyclicality. Cooling cycles reflect cosmic phenomena, while periods of global warming see an activation of water and carbon-based life forms and intensified competition for the earth's limited resources, affecting the giant insects and the giant reptiles, and equally the human race. In the case of the human race, the result of the intensified competition is globalized war and conflict, coups d'état, and terrorism. Table 9.3 shown, crop failures in cool periods led to the large migration of Germanic tribes and the great movement of the Huns, which was a chain reaction to poor harvests among nomadic tribes. Such territorial struggles over food resources, or over energy resources such as coal or petroleum, could pave the way to another world war. Both world wars were expected at the time to be short-lived conflicts, but, as is now known, events took an unexpected turn and an unforeseen situation developed.

Human-made disaster arises from and simultaneously with the risk of natural disaster in a scenario termed 'crisis synchronicity'. The accumulation of human-made disasters may lead to natural disaster and catastrophe in the near future, for instance through the polar ozone hole caused by the accumulation of CFC gas from refrigerators and hairspray. As history indicates, disaster involves both natural and human elements in a mutual chain reaction with the synchronicity phenomenon. The result has been a repeated cycle of destruction and creation which is counter-intuitive and defies human understanding. Interestingly as seen in Table 9.3, the global religions of Christianity, Islam, and Buddhism came into being at the beginning of a period of global warming, and this transition from a cool to a warm period with plentiful food supply and a growing population coincided with the outbreak of multiple conflicts. The conclusions might be that in cool periods the human race is preoccupied with securing food and is disinclined to engage in fighting amid the glacial conditions, and that large migrations take place in glacial periods with long-lasting food shortage. Taking a super-historical view, it would seem that the rise of religions and the outbreak of wars start in eras of global warming, while racial migration, internal conflicts, and state collapse from poor harvests occur in eras of global cooling. For reference see Table 9.3 Translational review between earth and human history.

If the present global warming is to be understood as a human-made disaster, then attention needs to be given not only to the incubating effect of CO_2 but also to the world's 441 nuclear power plants with an average capacity of 100 Mwe, each of which discharges water heated by 7 °C (70–100 ton/sec) for a period of approximately 40 years, leading to the phenomenon of ocean warming. It cannot be ruled out that the human race is thereby raising the risk of disaster. Changes in the global environment in the form of the warming-related climate crisis and abnormal weather patterns influence the cycles of convection and circulation in the atmosphere and the water sphere, inducing phenomena such as super-typhoons, heat waves, and extreme cold. These may also prevent heat release and circulation to adjust to global thermal differences and restore energy equilibrium, affecting even the activity of the earth's surface and the earth's crust (plate movement due to mantle convection). It is still unclear whether the great changes in the climate of today are dependent on cyclical solar activity or perhaps due to a change in the gravitational balance following the solar system planetary alignment of 2012. Nevertheless, the totality of human activity cannot be excluded as at least a factor in triggering disaster events. Human history demonstrates that disasters are not transitory events but have the ability to become chronic in character. In the face of the threat to the continued existence of the human race, I therefore felt the need to sound the alarm from the disaster-prone country of Japan, where we have frequent experience of earthquakes, tidal waves, volcanoes, typhoons, torrential rains, floods, landslides, and other natural disasters. In attempting this book, intended as a warning message to readers transcending location and generation, I wished to point to the potential for such natural disasters to join forces with human-made disaster and cause global devastation.

Atmospheric global warming and hydrospheric ocean warming (the boiling globe phenomenon) are explained by crisis sciences based on interdisciplinary within *transdisciplinary paradigms*.

9.3 Crisis Sciences as Transdisciplinary Paradigms

The aim of this book is to make a contribution to humanity by alerting the world to the gradually 'spreading unsafety' that faces us in the near future by presenting examples such as the Fukushima nuclear power disaster (radioactive contamination), environmental hormones (chemical substance contamination), the continuing series of railway accidents, viral infection pandemics, and the nuclear-heated ocean warming which is a factor in global warming. The idea is, firstly, to consider the natural disasters and unnatural disasters likely to confront us in the future by reviewing world history; secondly, to prevent the major misfortune of catastrophes combining natural with human-made disaster; and thirdly, for the next generations to explore the potential for realizing sustainability through 'citizen power for eco-civilization' as a preventive social function to counter the collusive relationship of business administration and public policy under modern capitalism.

This book consists of three parts encompassing a total of nine Chapters. Part I, Disaster Chain, begins with Chap. 1, Carbonized Terra, which traces the history of the earth and humankind to establish an understanding of the contemporary problem of unsafety; Chap. 2, The Fukushima Nuclear Catastrophe, serves as an example of the interconnectedness of natural disaster and human-made disaster; Chap. 3, Environmental Hormones, investigates the damage caused by the accumulation of chemical substances. Part II, Organizational Accidents, opens with Chap. 4, The JR West Accident, which focuses on a 2005 railway accident to present an example combining human error with system error; Chap. 5, Sociobiological Hazard, points to the northward and southward spread of contagious disease pandemics; Chap. 6, Boiling Globe, raises the issue of ocean warming caused by thermal effluents from the cooling process of the world's 441 nuclear reactors and presents a case study of the criticality accident at JCO's Tōkai nuclear-fuel plant. Part III, Science of Crises, begins with Chap. 7, Escape from Disaster, which applies the academic theories of pioneering scientists to explain the mechanism whereby a zone of indifference intervenes in the cognitive human process around disaster and accident, causing us to seek mental escape from risk and crisis; Chap. 8, Crisis Sciences for Sustainability: Limits of Management and Policy, warns of the collusive relationship between public policy and private management which threatens public safety and advocates the potential of an eco-civilization based on citizen power as a corrective social function; Chap. 9, Remaking Eco-civilization, traces humankind's journey to present a historical review of the contemporary unsafety which is the book's main theme and place it in context. To sum up, the book reexamines the danger of the risk latent in a market society leading to a collusive relationship between the normally opposing functions of organizational management and public policy, which may then become the source of crisis with 'moral destruction', and explores the possibility of the citizenship paradigm shift from capitalism to sustainism.

The book seeks to investigate examples of the organizational disasters and accidents which are a negative aspect of human cooperation in contemporary society and to think about how social systems can be designed to avoid their recurrence in the future. The world is full of needless disasters and accidents and instances of unsafety that should have been prevented, which more and more frequently plunge victims' families and loved ones into the greatest sorrow. Perhaps the time has come for those of us with the will to prevent the looming crisis of the near future to fight to assert citizen power over the state and companies and organizations. Today, the power residing in the state and organizational authority acts with increasing oppressiveness against the individual. A reaction based on eco-citizenship is the only social function able to prevent the unsafety that tends to be masked by organizational authority and state power. The intention of this book is to highlight the inescapable need for 'sustainability within eco-civilization' as a lesson proceeding from the combined disaster of the Great East Japan Earthquake (natural disaster) and the Fukushima nuclear accident (human-made disaster).

When one investigates and thinks about the issues facing contemporary society, one notices that they consist of the mixed blessings of the organized society created by modern rationalism. Part of the cause of this is that the methods of contemporary science are biased toward making microscopic distinctions relating to partial phenomena and analyzing related functions and structures from the limited approach of all normative individual sciences. While this method does have the ability to provide partial explanations for limited phenomena, it is not suited to elucidating overall phenomena. This is the problem of the 'subdivided sciences' to which G. Bateson refers. It is true that contemporary rationality has brought about economic development and a society of plenty through the pursuit of material wealth, but the negative side of the balance cannot be ignored. One example is global environmental pollution, as seen, for instance in the issue of environmental hormones, whose effects will impact not only contemporary society but also successive future generations. Meanwhile, the disasters and accidents that have occurred frequently in recent years have pointed up the limitations of the contemporary social system based on the logic of growth.

Another reason is that the design approach to knowledge and technology in contemporary society is the 'logic of the strong' centered on growth and development, which has marginalized and excluded the grieving and complaining weak. The negative aspects of human cooperation resulting from this are too numerous to recount, but include an increase in global environmental problems and unnatural disasters. Going forward, toward the realization of a sustainable society, the need has surely arisen to explore a preventive science that focuses on issues of safety. In modern capitalism, structures of collusion between business administration and politicians have also had a range of damaging effects. This collusion between business management and public policy, which should normally be opposing forces, leaves no option as a preventive function other than citizenship. The accelerating drive toward the 'market society and worship of money' under the capitalist system is bringing us to the 'limits of morality'. Preventive social functions would be similar to the parasympathetic nerves that control sympathetic nerves in living organisms, and we have now entered an age where they need to be engineered into social organizations.

The significance of the environmental management and disaster management (S. Bennett) of recent years is that it marks the advent of an age where, as it were, the sympathetic nerve for growth has been replaced by a parasympathetic nerve for prevention. The various social systems and control technologies which have developed in the organized society of today have expanded their applications rapidly in line with systems of information and communications technology, but have not contributed greatly to the prevention of unsafety at the global level. In historical terms also, individual scientific disciplines limited to specific situations are waning in their power to prevent the major misfortunes of human-made and natural disaster. As an individual citizen of Japan, the country where the Fukushima nuclear disaster occurred, I have a message to communicate to the world. That is, that the chain of unsought consequences that are the cumulative result of human negative cooperation can combine with natural disaster and lead to irrecoverable catastrophe. My reaction to global warming and other global environmental issues is to launch a global debate from Japan on the combination of human-made and natural disasters that result from accumulated negative cooperation and on the prevention of catastrophe for 'eco-civilization'.

Until now human science has been focused on matters of benefit to humankind and has been a study exclusively of fortunate circumstances, but we are now in an age where we need to include misfortune as a subject of scientific study. This should be clear from the examples given in Chaps. 2, 3, 4, 5, and 6 of this book. We need to think not of immediate gain or loss but of the moral aspect. A favorite book of businesspeople around the world is the *Book of Five Rings* by Musashi Miyamoto (a legendary swordsman of the 16th and 17th centuries). In his theory of the rings of elements, he advises humanity to "think lightly of oneself and deeply of the world" and "not envy another's good or evil." For the future of humanity, it can be concluded that the truth lies in the reality of misfortunes represented by disasters and accidents, namely, *unsafety*.

For the sake of the sustainability of the next generation, crisis sciences are rethinking the existing processes of decision-making at the individual, organizational, and social level.

This book examines accidents and disasters in the modern era, clarifies the mechanisms involved using real-life examples from Japan, and asks the question of how we can check the underlying pathology and threat of systemic breakdown. In recent years, disasters causing enormous misery have occurred across the globe at a frequent rate, while human-made organizational accidents have also inexorably grown in scale, presenting an urgent issue in all nations for individuals, organizations, regions, and the state. The objective of this publication is to explore potential management approaches (parasympathetic systems) and policies to limit the globally expanding problem of human-made disasters, drawing retrospectively on real-life examples from the Japan of recent years.

The role of the book is to contribute to the resolution of a range of emerging problems, from the aging of vital infrastructures for the supply of water, gas, and electricity to the breakdown of healthcare, pensions, and other social systems. One factor in the Fukushima nuclear catastrophe, which followed in the wake of the earthquake and tsunami disaster that struck eastern Japan in 2011, was the latent deterioration and aging of systems at all levels from the physical to the social, leading through 'chain reaction to unsought consequences' that no one foresaw, as explained in *The Functions of the Executive* (C. I. Barnard 1938). Here, the aging of the nuclear reactor system, the breakdown of safety management, and inappropriate instructions from the regulatory authorities combined to create a 'threefold disaster', in which technological, organizational, and governmental dysfunction has been diagnosed as reflecting a systems pathology infecting all levels.

Previous studies in the field of organizational accidents and human-made disasters have followed a well-known academic trajectory marked by works such as *Our Stolen Future* (T. Colborn 1996), *Remaking Eden* (L. Silver 1997), *Managing the Risks of Organizational Accidents* (J. Reason 1997), *World on the Edge* (L.R. Brown 2011), and *Collapse: How Societies Choose to Fail or Succeed* (J. Diamond 2011). Nevertheless, the academic discipline has yet to establish a sufficient identity. The historical development of scientific study in the fields of crisis management, risk cognition, safety policy, and reliability has been limited to specific disaster regions and countries. Since no widely shared history of disaster has been elaborated, this area can be viewed as a new global science or revolution.

As in America, Western Europe and Japan, trends in the BRICs nations (China, India, Brazil, and Russia) and other emerging economies suggest the risk of a global disaster escalating in the manner of a domino effect, as seen when inadequate safety management leads to the release of radioactive contamination, environmental hormones, or biohazards. Whether physical or social, whether process or reality (A. N. Whitehead), and whether micro or macro in scale, disasters share a common essence, which means that there is social value in establishing 'factual premises', as proposed by *Administrative Behavior* (H. A. Simon), on the basis of these frequent phenomena. These 'factual and valuable premises' will be significant as 'scientific paradigms' as in *The Structure of Scientific Revolutions* (T. Kuhn) and can be seen as occupying a perspective that integrates theoretical, empirical, and pragmatic science. Indeed, this book's approach can be considered as an interdisciplinary science based on this integration of normative methods.

In its structure, this book distinguishes between physical disasters (typhoons, earthquakes, tsunamis, floods), chemical disasters (environmental hormones, polychlorinated biphenyls, dioxins, PM2.5), and biological disasters (biohazards, virus pandemics, bio-manipulation), and is based on an interdisciplinary approach to preventing disaster through examination of the process of mutual repercussion that is at the genesis of disaster. In view of current needs, with the number of people and regions affected by disaster rapidly increasing, the growth of research in this area can be seen as science fulfilling its social mission in response to a tacit societal desire for a future where safety is guaranteed. The book's aim is thus not merely to launch a critique of business and government, but rather to contribute to achieving a sustainable society in the near future.

9.4 Afterword

The inspiration for this research arose from a number of unsettling personal encounters with unsafety. Tracing them back in reverse order, I was affected by the EU refugee crisis and global terrorism; in December 2015, in the German town of Fussen, I was forced by the hotel manager to leave my hotel in the early morning without breakfast because of a misunderstanding and felt unsafe due to the tense situation in the town caused by the EU refugee crisis; the Fukushima Nuclear Disaster and the Great East Japan Earthquake of 2011; the Thai coup d'état of 2007 (during transit at Bangkok airport on my way to a volunteering event at Mt. Everest); an Alaskan wild grizzly scare in a 300 km-long canoe trip on the Yukon River in 2007; the JR Fukuchiyama Line derailment accident of 2005 (to which a student from one of my classes fell victim); an extreme cold snap in a mountaineering trip to the Matterhorn in August 2003, on the contrary an extreme heat wave on Gran Paradiso and rock fall caused by glacier melt at Mont Blanc in 2003; the SARS pandemic that swept across Asia in 2003 (during which I was delayed by a major disinfection carried out at Kansai airport while on my overseas research posting to Prof. M. von Zedtwitz at the IMD (Institute for Management Development) in Lausanne; the coup d'état by Peru's Fujimori government in 1999 (during which I experienced an enforced stay at Machu Picchu by the rebel troops); and the Kobe earthquake of early morning in 1995. Unfortunately, it was thus the unsafety that I experienced personally that proved sufficient motivation for the publication of this book.

Bibliography

ARAIC (2006), The Aircraft and Railway Accident Investigation Commission, *Jijitsu ni kansuru hōkokusho-an no Ikenchōshukai* [*Draft Report Concerning Fukuchiyama Line Derailment Accident*], JTSB, 2006.

ARAIC (2007), The Aircraft and Railway Accident Investigation Commission, *Tetsudō jiko chōsa hōkokusho Nishinihon ryokaku tetsudō kabushikigaisha Fukuchiyama ~ Amagasakikan resshadassenjiko,* [*Fukuchiyama Line Derailment Accident Investigation Report*], JTSB, 2007.

Abe, H., *AIDS to wa Nanika [What is AIDS?]*, NHK Publishing, 1986, p. 28.

Abe, S. (eds), *Public Transport is Dangerous: Deregulation and Work Overload*, Iwanami Shoten, 2005, p. 11.

Acemoglu, D. and Robinson, A., *Why Nations Fail: The Origin of Power, Prosperity and Poverty*, Brockman, Inc., 2012.

Ackoff, R.L., "On Misdirected Systems" (Read by Mike Jackson), *World Congress of the Systems Sciences*, International Society for the Systems Sciences, Toronto, 2000, p. 6.

Agency for Natural Resources and Energy, *Genshiryokuhatsuden no haishisochi wo meguru kaikeiseido no kadai to ronten [Issues and Questions Concerning Accounting Systems for the Measures around Nuclear power Decommissioning]*, 2013.

Andrews, K., "Practice in Management", *Harvard Business Review*, May–Jun.1987.

Argyris, C. and Schön, D., *Organizational Learning: A Theory of Action Perspective*, Addison-Wesley, 1978.

Argyris, C., *Integrating the Individual and the Organization*, Transaction Publishers, 1974, 1990.

Argyris, C., *Knowledge for Action*, Jossey-Bass Publishers, 1993.

Arrow, K.J., *The Limits of Organization*, Norton & Company, 1974.

Ashby, W.R. *An Introduction to Cybernetics*, London: Chapman & Hall, 1956.

Ashby, W.R., *Design for a Brain*, John Wiley & Sons, 1952.

Atsuji, S. and Cook, N.D., "Fukushima Nuclear Catastrophe 3.11: System Pathology of Social Organizations", *Proceedings of the 55th World Congress of the ISSS,* International Society for the Systems Sciences, Hull UK, 2011, p.82. (Full paper pp.1–14; URL: http://journals.isss.org/index.php/proceedings55th/article/view/1637)

Atsuji, S. et al., "Sustainable Decision-Making Following the Fukushima Nuclear Catastrophe", *Proceedings of the 11th World Congress of the IFSAM*; ISBN 978-1-907300-04-2, International Federation of Scholarly Associations of Management, Limerick Ireland, 2012, p.145. (Full paper: the World Congress of the IFSAM, CD-ROM; ISBN 978-1-907300-05-9, pp.1–15)

Atsuji, S. et al., "Un-safety: Systems Pathology of the Fukushima Nuclear Catastrophe", *Proceedings of the 57th World Congress of the ISSS*, International Society for the Systems

S. Atsuji, *Unsafety*, Translational Systems Sciences 7,
DOI 10.1007/978-4-431-55924-5

Sciences, Hai-phong Vietnam, 2013, p.68. (Full paper pp.1–16; URL: http://journals.isss.org/index.php/proceedings57th/article/view/2017)

Atsuji, S. et al., "Our Stolen Safety: Systems Pathology following the Fukushima Nuclear Disaster and the JR Accident", *Proceedings of the 57th World Congress of the ISSS*, International Society for the Systems Sciences, Hai-phong Vietnam, 2013, p.68. (Full paper pp.1–21; URL: http://journals.isss.org/index.php/proceedings57th/article/view/2016)

Atsuji, S. et al., "The Fukushima Catastrophe seen as a Malfunction of Organizational Systems: Nuclear Fade-out and Hazard Maps for the World", *General Systems Bulletin XXXXI*, International Society for the Systems Sciences, 2012, vol. 41, pp.26–30. (URL: http://isss.org/world/system/files/files/2012%20GSB.pdf)

Atsuji, S. et al., "Disaster Management for Boiling Globe by World's 441 Nuclear-heated Water", *Program and Abstracts of the ISSS*, International Society for the Systems Sciences, Berlin Germany, 2015, pp.134. (Full paper pp.1–15; URL: http://journals.isss.org/index.php/proceedings59th/author/submission/2587)

Atsuji, S. et al., "Environmental Sustainability in Fukushima Nuclear Crisis", *Proceedings of the 12th World Congress of the IFSAM*, International Federation of Scholarly Associations of Management, Tokyo, 2014, p.76. Expanded abstract: CD-ROM, pp.1–15. (URL: http://www2.convention.co.jp/ifsam2014/program.html)

Atsuji, S. et al., "Organizational Disaster by Implicit Systems Errors: A Case Study of the JR West Accident", *Proceedings of the 11th World Congress of the IFSAM*; ISBN 978-1-907300-04-2, International Federation of Scholarly Associations of Management, Limerick Ireland, 2012, p.154. (Full paper: CD-ROM; ISBN 978-1-907300-05-9, pp.1–12)

Atsuji, S. et al., *JR Accident 4.25*, Kansai University, 2007.

Atsuji, S. and Senba, S., "Denkijigyōjiyūka to shakaikōkyōsei", in *Kōgyō keiei kenkyū* [*Industrial Management Research*], Kōgyō Keiei Kenkyū Gakkai, 2002, vol. 16, pp. 22–26.

Atsuji, S., *Management Policy for Organizational Disaster*, Doshisha University, 2003, pp. 1–200.

Atsuji, S., *Soshiki Kettei no Jyouhou Katei [The Informatics of Organizational Decision-making]*, Japan Academy of Business Administration, 2000, pp. 128–135.

Atsuji, S., *Soshiki Kettei no Kagaku [The Science of Organizational Decision-making]*, Kansai University, 1999, pp.1–240.

Atsuji, S., "ISO Global Policy toward Environmental Management", *Economic & Political Studies Series No. 135*, Institute of Economic and Political Studies, 2003, pp. 1–18.

Atsuji, S., "What is the Basis of Complexity?: Cybernetics on the World Congress of the Systems Sciences", *Social and Economic Systems Studies*, The Japan Association for Social and Economic Systems Studies, 2001, No. 20, pp. 149–158.

Awano, M., "Ano hi Tōkaimura de nani ga okotta ka" [What Happened That Day at Tokaimura?], Nanatsumori Syokan, 2001, p. 103.

Ayer, A.J., *The Foundations of Empirical Knowledge*, Macmillan, 1940.

Barnard, C.I., *Organization and Management*, Harvard University Press, 1948.

Barnard, C.I., *The Functions of the Executive*, Harvard University Press, 1938.

Bateson, G., *Steps to an Ecology of Mind*, Paladin, 1973.

Beck, U., *Risk Society: Towards a New Modernity*, Sage, 1992.

Beck, U., *World at Risk,* Polity Books, 2009.

Beer, S., "World Organization of Systems and Cybernetics (WOSC)", *World Congress of the Systems Sciences*, International Society for the Systems Sciences, Toronto, 2000, p.5.

Beer, S., *Brain for the Firm; 2nd edition*, John Wiley, 1981.

Bennett, S., *Innovative Thinking in Risk, Crisis, and Disaster Management*, Gower Publishing, 2012.

Berger, P.L. and Luckmann, T., *The Social Construction of Reality*, Doubleday, 1966.

Berle, A.A., and Means, G.C., *The Modern Corporation and Private Property*, Transaction Publishers, 1932.

Bertalanffy, L.V., *Das Biologische Weltbild*, A. Franke AG. Verlag, 1949.

Bertalanffy, L.V., *General System Theory*, George Braziller, 1968, pp.47–48.

Birkmann, J. (eds), *Measuring Vulnerability to Natural Hazards.* New York: United Nations University, 2006.
Bohm, D., *Fragmentation and Wholeness*, Van Leer Jerusalem, 1976.
Bohm, D., *Wholes and the Implicate Order*, Routledge & Kegan Paul, 1980.
Bohm, D. (eds.), *Science et Conscience*, Stock & France Culture, 1980.
Brown, H.I., *Perception; Theory and Commitments*, The University of Chicago, 1977.
Brown, L.R. et al., *Saving the Planet: How to Shape an Environmentally Sustainable Global Economy*, Worldwatch Institute, 1991.
Brown, L.R, et al., *State of the World 2000*, Norton, 2000.
Brown, L.R., et al., *Vital Signs: The Trends that Are Shaping Our Future*, Worldwatch Institute, 1992.
Brown, L.R., *Plan B 4.0: Mobilizing to Save Civilization*, W.W. Norton & Company, 2009.
Brown, L.R., *World on the Edge*, Earth Policy Institute, 2011.
Bruner, J.S., *In Search of Mind*, Harper & Row, 1983.
Buckley, W., *Sociology and Modern Systems Theory*, Prentice-Hall, 1967.
Cadbury, D., *The Feminization of Nature*, Penguin, 1997.
Capra, F., *The Tao of Physics*, Murray, 1975.
Carrel, A., *The Man Unknown*, Harper and Bros., 1935.
Carroll, G.R. and Hannan, M.T., *Organizations in Industry*, Oxford University Press, 1995.
Carroll, G.R., "On the Organizational Ecology of Chester I. Bernard", Williamson, O.E. (eds), *Organization Theory; from Chester Barnard to the Present and Beyond*, Oxford University Press, 1990, pp.56–71.
Carson, R., *Silent Spring*, Houghton Mifflin, 1962.
Cashore, B. and Vertinsky, I., "Policy Networks and Firm Behaviors", *Policy Sciences*, Vol. 33: No. 1., 2000, pp. 4–9.
Cassirer, E., *An Essay on Man; An Introduction to a Philosophy of Human Culture*, Yale University Press, 1944.
Chaisson, E., "Order, Flow and Complexity: A Precis on Cosmic Evolution", *Proceedings of World Congress of Systems Sciences*, International Society for the System Sciences, Toronto, 2000, pp. 22–23.
Chroust, G., "Regional Disasters and Systemic Reactions", *Program and Abstracts of the 55th Annual Meeting of ISSS,* International Society for the System Sciences, Hull UK, 2011, p. 97.
Chroust, G., "ICT Support for Disaster Management", In Doucek, P., Chroust, G. and Oskrdal, V. (eds.), *IDIMT ICT-Support for Complex Systems,* Trauner Verlag Linz, *vol.38*, 2012, pp. 13–23.
Chroust, G. and Sankaran, S., "Information and Communication in Disaster Response", *Program and Abstracts of the 57th Annual Meeting of ISSS,* International Society for the System Sciences, Hai-phong Vietnam, 2013, pp. 51–52.
Chroust, G. and Aumayr, G., "Reliable Communication – A Key to Disaster Response", *Program and Abstracts of the 57th Annual Meeting,* International Society for the Systems Sciences, Hai-phong Vietnam, 2013, pp. 54–55.
Citizen's Nuclear Information Center, *Kenshō Tōdengenpatsu toraburugakushi* [*Inspection: Concealment of Trouble at Tokyo Electric's Nuclear Plants*], Iwanami Booklet No. 582, 2002.
Civil Accident Investigation Committee.
Colborn, T., Dumanoski, D. and Myers, J. P., *Our Stolen Future: Are We Threatening Our Fertility, Intelligence, and Survival?*, *Our Stolen Future*, A Dutton Book, 1996. (http://www.wwf.or.jp/lib/lecture/colborn001028/career.htm)
Cook, N. D. "Artefact or network evolution?" *Nature Publishing Group*, Vol. 374 (6520), Mar 1995, 313.
Cook, N.D. and Atsuji, S., "Fukushima Nuclear Catastrophe 3.11: System Pathology of Social Organizations", *Journal of the International Society for the Systems Sciences*, 2011. (Full Paper; http://journals.isss.org/index.php/proceedings55th/article/view/1637)

Cribb, J., *The Coming Famine: The Global Food Crisis and What We Can Do to Avoid It*, University of California Press, 2010.
Cyert, R.M. and March, J.G., *A Behavioral Theory of the Firm*, Prentice-Hall, 1963.
Davis, S.T., *Managing Corporate Culture*, Harper & Row, 1984.
Deguchi, H., *Economics as an Agent-Based Complex System: Toward Agent-Based Social Systems Sciences*, Springer, 2004.
Diamond, J.M., *Collapse: How Societies Choose to Fail or Succeed*, Viking Penguin, 2005.
Diamond, J.M., *Guns, Germs, And Steel: The Fates of Human Societies*, W. W. Norton & Company, 1997.
Diamond, J.M., *The World Until Yesterday: What Can We Learn from Traditional Societies?*, Viking, 2012.
Dowling, P., A Speech of IFSAM Presidential, International Federation of Scholarly Associations of Management, Paris, 2006.
Drucker, P.F., *Innovation and Entrepreneurship*, Harper Collins Publishers, 1985.
Drucker, P.F., *The Essential Drucker: The Best of Sixty Years of Peter Drucker's Essential Writings on Management*, Harper, 2001.
Eccles, J.C., *Facing Reality*, Springer Verlag, 1970.
Farrell, A., "Sustainability and Decision-Making", *Policy Studies Review*, Vol. 16: Nos.3/4 1999, pp. 37–43.
Fayol, H., *Administration industrielle et generale*, Dunod et Pinat, 1917.
Feigenbaum, E.A. and MaCorduck, P., *The Fifth Generation*, Signet, 1983.
Feldheim in Germany, *Available* at: http://www.neue–energien–forum–feldheim.de/. *Accessed* April, 2014.
Feldman, E., and Bayer, R. (eds), *Blood Feuds: AIDs, Blood, and the Politics of Medical Disaster*, Oxford University Press, 1999.
Fennell, D., *Investigation into the King's Cross Underground Fire*, Department of Transport, London: HMSO, 1988.
Fink, G. et al., "Survival by Subversion in Former Socialist Economics: Tacit Knowledge Exchange at the Workplace", International Federation of Scholarly Associations of Management, Berlin, 2006, p.32.
Free, R., *The Role of Procedural Violations in Railway Accidents* (PhD Thesis), University of Manchester, 1994.
Fromm, E., *Escape from Freedom*, Rinehart and Company, 1941.
Fromm, E., *Man for Himself: An Enquiry into the Psychology of Ethics*, Rinehart and Company, 1947.
Fukushima Genpatsu Jiko Dokuritsu Kenshō Iinkai [Fukushima Nuclear Plant Accident Independent Inspection Committee, *Chōsa/kenshō hōkokusho* [*Investigation/Inspection Report*].
Fuller, R.B., *Operating Manual for Spaceship Earth*, Southern Illinois University Press, 1969.
Gardner, H., *The Mind's New Science*, Basic Books, 1985.
Garvin, D.A., *Learning in Action: A Guide to Putting the Learning Organization to Work*, Harvard Business School Press, 2000.
Gassmann, O. and Zedtwitz, M.von., "Trends and determinants of managing virtual R&D teams", *R&D Management* 33, 3, Blanckwell Publishing Ltd, 2003.
Gassmann, O., "Research vs. Development: Towards a New Framework of R&D Internationalization", International Federation of Scholarly Associations of Management, Berlin, 2003, pp. 243–262.
Gazzaniga, M.S., *The Social Brain*, Basic Books, 1985.
Gioia, D.A., "The State of the Art in Organizational Social Cognition: A Personal View", Sims, H. P., Gioia, D.A., and Associates, *The Thinking Organization; Dynamics of Organizational Social Cognition*, Jossey-Bass Publishers, 1986.
Glendon, A.I. and McKenna, E.F. *Human Safety and Risk Management*, London: Chapman & Hall, 1995.
Gore, A. and Bosquet, M., *Ecologie et Politique*, Ed. Galil, 1975.

Gore, A., *Earth in Balance*, Houghton Mifflin, 1992.
Gore, A., *The Future: Six Drivers of Global Change*, Random House, 2013.
Gurr, T.R., *Why Men Rebel*, Princeton University Press, 1970.
Hauser, P. (eds), *'Thyroid Diseases of Infancy and Childhood: Effects on Behavior and Intellectual Development'*, American Psychiatric Press, 1999. (http://www.health–station.com/topic92.html)
Heims, S.J., *John Von Neumann and Norbert Wiener*, MIT Press, 1980.
Heinrich, H.W., *Industrial Accident Prevention – A Scientific Approach*, McGraw-Hill, 1931.
Hidden, A., "Investigation into the Clapham Junction Railway Accident", *Department of Transport London HMSO*, 1989.
Hioki, K., *Ichiba no Gyakushu [Market's Counterattack: Restoration of Personal Communications]*, Taishukan, 2002.
Hioki, K., *Principles of Management,[Keieigaku Genri]*, Economist, 2000.
Hioki, K and Nakamaki, H., *Keiei Jinruigaku Kotohajime [Beginning; Management Anthropology]*, Touho, 1997.
Hiromatsu, W., *Mono, Koto, Kotoba*, Keisou-shobo, 1979.
Hochberg, J.E., *Perception*, 2nd edition., Prentice-Hall, 1978.
Hofstadter, D.R., *Goedel, Escher, Bach*, Basic Books, 1979.
Humphrey, N., *The Inner Eye*, Faber and Faber, 1986.
Husserl, E., *L'Origine de La Geomtrie*, Presses Universitaires de France, 1962.
IAEA, "Inventory of Radioactive Waste Disposals at Sea", IAEA-TECDOC-1105, 1999.
IPCC: "Intergovernmental Panel on Climate Change", *Climate Change 2013: The Physical Science Basis*, 2013.
Iino, H., "Soshikidōtoku to soshikibunka", *Barnard: Gendaishakai to Soshikimondai [Organizational Ethics and Organizational Culture, Barnard Modern Society and Organizational Problems]*, Bunshindō, 1986, pp. 286–290.
Iino, H., *Barnard Kenkyu*, Bunshindo, 1983.
Iino, H., *Barnard Soshiki-ron Kenkyu*, Bunshindo, 1992.
Isenberg, D.J., "Drugs and Drama; The Effects of Two Dramatic Events in Pharmaceutical Company on Manager's Cognitions", Working Paper #83–55, Boston MA, Harvard Business School, 1983.
Isenberg, D.J., "How Senior Managers Think", Henry, J., *Creative Management; The Open University 1991*, Sage Publications, 1991, pp.43–56.
Isenberg, D.J., "Thinking and Managing; A Verbal Protocol Analysis of Managerial Problem Solving", *Academy of Management Journal*, Dec.1986, p. 775.
Isenberg, D.J., "The Structure and Process of Understanding; Implications of Managerial Action", Sims, H.P., Gioia,D.A., and Associates, *The Thinking Organization; Dynamics of Organizational Social Cognition*, Jossey-Bass Publishers, 1986.
JACO, *ISO14000 Kankyō management system [ISO 14001 Environmental Management System (Bilingual)]*, Gyōsei, 2001.
JCO "Rinkai Jiko Sōgō Hyoka Kaigi", (Meeting Presenting Overall Assessments of the JCO Criticality Accident), *JCO rinkai jiko to Nihon no genshiryoku gyōsei. [The JCO Criticality Accident and Japan's Nuclear power Administration]*, Nanatsumori Shokan, 2000.
JR West, *JR West in Figs. 2008*, JR West, p. 28.
Jacobson, J.L. and Jacobson, S.W., "Intellectual Impairment in Children Exposed to Polychlorinated Biphenyls in Utero", *New England Journal of Medicine*, September 1996.
Japan Nuclear Energy Safety Organization, "Jishin Reberu 2PSA no Kaiseki (BWR)" [Level 2PSA Earthquake Analysis], October 2010, pp. 4–7.
Japan Society of Geology (eds), S. Kiyokawa, T. Itou, M. Ikehara and T. Onoue, *Super Time Scale of the Earth,* Iwanami, 2014.
Kanuma, K., *Yakugai AIDS Saikō [Rethinking AIDS]*, Kadensha, 1998, p. 21.
Kataoka, N., "Keiei Kouzou Bunseki –Jigyou, Kigyou, Keiei", Yamamoto and Katou, *Keieigaku Genron [Principle of Management]*, Bunshindo, 1982.
Kataoka, N., *Gendai-kigyo no Shoyuu to Shihai*, Hakutou-shobou, 1992.

Kataoka, N., *Gendai-kigyo Shakai ni-okeru Kojin no Jiritsu*, Bunshindo, 2004.

Kato, K., *Barnard and Henderson*, Bunshindo, 1996.

Kato, H., *Kankyō to rinri: Shizen to ningen no kyōsei wo motomete* [*Environment and Ethics: Towards the Coexistence of Nature and Humanity*], Yūhikaku, 1998, pp. 1–19.

Kawanaka, N., *Seisaku Kettei to Shakai Riron, [Policy Decision-making and Social Theory]*, Ryōsho Fukyūkai, 1984, pp. 16–19.

Kemeny, J., *The Need for Change: The Legacy of TMI. Report of the Presidential Commission at Three Mile Island*, New York: Pergamon, 1979.

Kessler, D. J. and Cour–Palais, B.G., "Collision Frequency of Artificial Satellites: The Creation of a Debris Belt", *Journal of Geophysical Research* 83, 1978.

Kijima, K., "Translational and Trans-disciplinary Approach to Service Systems", Kijima, K.(edit), *Service Systems Science; Translational Systems Sciences 2*, Springer, 2015, pp.40–42.

Kineman, J.J., "Crisis sciences for sustainability", *Program and Abstracts of the ISSS*, International Society for the Systems Sciences, Berlin, 2015, pp.166–167.

Koide, H., *Genpatsu no uso [The Lies about Nuclear Power]*, Fusōsha, 2011, p. 119.

Kokkai jiko chōhōkokusho [*Accident Report to the Diet*], *Tōkyō Denryoku Fukushima Genshiryoku Hatsudensho Jiko Chōsa Iinkai [Tokyo Electric Fukushima Nuclear Plant Accident Commission]*, Tokuma Shoten, 2012.

Koontz, H., O'Donnell, C., and Weihrich, H., *Management: 7th edition*, McGraw-Hill, 1983.

Kotter, J.P., *Organizational Dynamics: Diagnosis and Intervention*, Addison-Wesley, 1987.

Kouno, T., *Keiei-Soshiki no Kagaku to Ginou to Rinri*, Chikura-shobou, 2003.

Koutou, I., *Keiei-tetsugaku karano Sekinin no Kenkyu*, Bunshindo, 2013.

Krauss, L.M., *A Universe from Nothing: Why There is Something Rather than Nothing*, Simon & Schuster, 2012.

Kuhn, T.S., *The Structure of Scientific Revolutions*, The University of Chicago, 1977.

Langer, S.K., *Philosophy in a New Key: A Study in the Symbolism of Reason, Rite, and Art*, Harvard University Press, 1942.

Langer, S.K., *Problems of Art: Ten Philosophical Lectures*, Charles Scribner's Sons, 1957.

Langton, P., *The Black Triangle*: *Environmental Policy in an International Context*, The Open University, BBC–EPIC, 1999.

Laszlo, E., "The Evolutionary Challenge of the Third Millennium", *World Congress of the Systems Sciences*, International Society for the Systems Sciences, Toronto, 2000, pp.25–26.

Leiserowitz, A., "Climate change risk perception and policy preferences: The role of effect, imagery, and values", *Climate Change* 77: 45–72. 2006.

Levi-Strauss, C., *Tristes Tropiques*, Plon, 1976.

Life AIDS Project (http://www.lap.jp/lap2/data/yakugain.html).

Likert, R., *New Patterns of Management*, Garland, 1967.

Lindsay, P.H. and Norman, D.A., *Human Information Processing; An Introduction to Psychology*, 2nd edition., Academic Press, 1977.

Litwin, G.H. and Stringer, R.A., *Motivation and Organizational Climate*, Harvard Business School Press, 1986.

Lloyd, C., *What on Earth Happened?: The Complete Story of the Planet, Life and People from the Big Bang to the Present Day*, Bloomsbury Publishing, 2012.

Lovins, A.B. et al., *Saiseikanō Energy ga Hiraku Mirai [A Future Built on Renewable Energy]*, Iwanami, 2013.

Luhmann, N., *Soziologische Aufklarung*, Opladen, 1974.

Mainichi Shimbun Shakaibu, *Yakugai AIDS Ubawareta Mirai [Medically Induced AIDS: Stolen Future]*, Mainichi Shimbunsha, 1996.

Mainichi Shimbun, 23 September 2001.

March, J.G. and Cyert, R.M., *A Behavioral Theory of the Firm*, Prentice-Hall, 1963.

March, J.G. and Olsen, J.P., *Ambiguity and Choice in Organization,* Universitetsforlaget, 1976.

March, J.G. and Simon, H.A.,*Organizations*; 2nd edition, Blackwell Publishers, 1993 (1st edition,1953).

March, J.G., *Decisions and Organizations*, Basil Blackwell, 1988.
Masai, Y., (ed.), *Ima ga wakaru jidai ga wakaru nihon chizu [Understanding the Present, Understanding the Period: Map of Japan]*, 2010 edition, Seibido Shuppan, 2009, p. 93.
'Masdar City' in UAE, *Available* at: http://masdarcity.ae/en/. *Accessed* May, 2014.
Maslow, A.H., *Motivation and Personality*, 2nd edition, Harper & Row, 1954.
Maturana, H.R. and Varela, F.J., *Autopoiesis and Cognition: The Realization of the Living*, D. Reidel Publishing Company, 1980.
Maturana, H.R., "Conservation and Change", *World Congress of the Systems Sciences*, International Society for the Systems Sciences, Toronto, 2000, p. 26.
Maturana, H.R., "Ethics as a Human Action—Ethics Belongs to the Emotional Domain", *Proceedings of the 44th Congress of the Systems Sciences,* International Society for the Systems Sciences, Toronto, 2000, p. 91.
Mayo Elton, *The Human Problems of an Industrial Civilization*, Macmillan, 1933.
McFarlan, W.F. and Mckenney, J.L., "The Information Archipelago; Maps and Bridges", *Harvard Business Review*, Sep-Oct. 1982.
McMillan, C.J., *The Japanese Industrial System,* De Gruyter, 1996.
McNeill, W.H., *A World History*, Oxford University Press, 1998.
McDonough, A.M., *Information Economics and Management Systems*, McGraw-Hill, 1963.
Meadows, D.H., Meadows, L.M., and Randers, J., *Beyond the Limits*, Chelsea Green, 1992.
Meadows, D.H., Randers, J. and Meadows, D.L, *Limits to Growth: The 30-Year Update*, Chelsea Green Publishing, 2004.
Meister, D., "Human error in man-machine systems", in Brown, S. and Martin, J. (eds), *Human Aspects of Man-made Systems,*
Merleau-Ponty, M., *Le Visible et L'invisible: Suivi de notes de travail*, Gillimard, 1964.
Midgley, G. (eds), *Systems Thinking, Volumes I–IV*, Sage Publications, 2003.
Midgley, G., "Systems Thinking for the 21st Century", *World Congress of the Systems Sciences*, International Society for the Systems Sciences, Toronto, 2000, p. 26–27.
Merleau-Ponty, M., *Le Visible et L'invisible*, Gallimard, 1964.
Merleau-Ponty, M., *Phenomenologie de la perception*, Gallimard, 1945.
Meteorological Research Institute of the Japan Meteorological Agency and the Advanced Earth Science and Technology Organization.
Miller, G.A., "The Magical Numbers Seven plus or minus Two; Some Limit on Our Capacity for Processing Information", *Psychological Review*, 1956, Vol.63, pp.81–97.
Ministry of Justice, *A guide to the Corporate Manslaughter and Corporate Homicide Act 2007*; *Available* at: (http://www.hseni.gov.uk/guidetomanslaughterhomicide07.pdf#search='organizational+Manslauguter'. *Accessed* May, 2014)
Ministry of Land, Infrastructure, Transport and Tourism, Railway Bureau website.
Mintzberg, H., *Mintzberg on Management: Inside Our Strange World of Organizations*, New York: The Free Press, 1989.
Minzberg, H., "Planning on the Left Side and Managing on the Right", *Harvard Business Review*, July 1976, pp.49–58.
Mitchell, T.R., Rediker, K.J., and Beach, L.R., "Image Theory and Organizational Decision Making", Sims, H.P., Gioia, D.A., and Associates, *The Thinking Organization; Dynamics of Organizational Social Cognition*, Jossey-Bass Publishers, 1986, pp.293–315.
Mito, T., *Zuihanteki-kekka: Kanri no Kakumei*, Bunshindo, 1995.
Monod, J., *Le Hasard et la Necessite*, Knopf, 1971.
Morley, M., Heraty N. and Collings, D. (eds.), *International Human Resource Management and International Assignments*, London: Palgrave Macmillan, 2006.
Morley, M., "A Speech of IFSAM Presidential", *Programme and Proceedings of the 11th World Congress of IFSAM,* International Federation of Scholarly Associations of Management, Limerick Ireland, 2012, pp. 7–8.
Morley, M., "On Coping, Competence and Capacity Building: The Nature and Role of Expatriate Cross-Cultural Training among Recently Internationalized Irish MNCs", *Program and*

Abstracts of the 12th World Congress of IFSAM, International Federation of Scholarly Associations of Management, Tokyo, 2014, pp.37–38.
Morris, I., *Why the West Rules – for Now: The Patterns of History, and What They Reveal About the Future*, Picador; 2011.
Murakami, Y., *Kindaikagaku wo koete*, Koudansha, 1986.
Murata, H., *Kanri no tetsugaku: Zentai to kō, sono hōhō to imi* [*Philosophy of Management: The Whole and the Part, Methods and Meanings*], Bunshindō, 1984.
Murata, H., *Jouho to Sisutemu no Tetsugaku*, Bunshindo, 1992.
Naisbitt, J., *Global Paradox*, Leigh, 1994.
NASA AIMS, *Available* at: http://www.nasa.gov/centers/ames/home/. (*Accessed* April, 2014)
National Energy Authority of Iceland, *Available* at: http://www.nea.is/. (*Accessed* May, 2014)
Neisser, U., *Cognition and Reality*, Freeman and Company, 1976.
Neisser, U., *Cognitive Psychology*, Prentice-Hall, 1967.
Nelson, H., "Continuing the Traditions of ISSS – The Reintegration of Sophia", *Proceedings of the 44th of the Systems Sciences,* International Society for the Systems Sciences, Toronto, 2000, p.36.
Neumann, J.V., *The Computer and the Brain*, Yale University, 1958.
News for the People in Japan, *Available* at: http://www.news-pj.net/diary/1001. (*Accessed* May, 2014)
NIKKEI, Lesson from the Crisis: Finding Light in the Storm, *The 11th Nikkei Global Management Forum*, 2010.
NHK ONLINE. *Available* at: http://www.nhk.or.jp/gendai/kiroku/detail_3471.html#marugotocheck. (*Accessed* May, 2014)
NHK Tōkaimura Rinkai Jiko Chōsahan *[Tokaimura Criticality Accident Investigation], Kuchiteitta inochi—Hibaku chiryō 83 nichikan no kiroku*, Shinchōsha, 2006.
NHK documentary wave, *Available* at: http://www.nhk.or.jp/documentary/. (*Accessed* April, 2014)
Nicholas, J.D., Pickett, G.B. and Spears, W.O., *The Joint and Combined Staff Officer's Manual*, Stackpole, 1959.
Nihon Kōgyō Shimbun publication, *Kakusareta genpatsu deeta* [*Concealed Nuclear-plant Data*], Fuji Sankei Group Nihon Kōgyō Shimbunsha, 2003, p. 14.
Nikkei Telecom 21, Newspaper article databases (http://edu21.nikkeimm.co.jp) (*Accessed* on November 13, 2006).
Nishida, K., *Zen no Kenkyu*, Iwanami-shoten, 1950 (1st edition 1911)
Nishioka, T., *Shijo Soshiki to Keiei-rinri*, Bunshindo, 1996.
Niwamoto, *Y., Barnard Keieigaku no Tenkai*, Bunshindo, 2006.
Niwamoto, Y., "Kindai kagakuron wo koete", *Osaka Shougyou Daigaku Ronshu* No.66, 1983, pp.119–131.
Nonaka, I., *Soshiki-Shinkaron*, Nihonkeizai-shinbunsha, 1985.
Norman, D.A., *Learning and Memory*, Freeman and Company, 1982.
Norman, D.A., *Models of Human Memory*, Academic Press, 1970.
Norman, D.A., *The Psychology of Everyday Things*, Basic Books, 1988.
Nuclear Safety Commission, Uranium [Kakō Kōjō Rinkai Jiko Chōsa Iinkai], "Uranium Kakō Kōjō Rinkaijiko Chōsa Iinkai Hōkoku"; [Uranium-Processing Factory Criticality Accident Investigation Committee], *Uranium-Processing Plant Criticality Accident Investigation Report,* 1999, III–1–(6).
Nuclear Safety Commission, "Kakunenryō shisetsu anzen shinsa kihon hōshin", Basic Measures for Safety Inspections of Nuclear–fuel Facilities; (http://nsc.jst.go.jp/anzen/sisin/sisin025/si025.html).
Nuclear and Industrial Safety Agency, "Genshiryokuhatsudensho no hoan katsudō hyōka (shikō) no kekka ni tsuite", [Regarding (trial) evaluation of activities to maintain nuclear plants], June 2010.
Odam, E.P., *Basic Ecology*, CBS College Publishing, 1983.

Odum, H.T.,"Systems Self Organization for the Prosperous Way Down", *Proceedings of the 44th Congress of the Systems Sciences,* International Society for the Systems Sciences, Toronto, 2000, p.23.

Osaka HIV Sosho Bengōdan, *Yakugai AIDS Kokusai Kaigi [Medically Induced AIDS International Conference]*, Sairyūsha, 1998, p. 161.

Oshima, K., *Saiseikanō enerugii no seijikeizaigaku* [*Political Economy of Renewable Energy*], Tōyō Keizai Shinpōsha, 2010, p. 080.

Ota, S., *Kigyou to Seisaku (Business and Policy)*, Minerva, 2013.

Oxenfeldt, A.R., Miller, D.W. and Dickinson, R.A., *A Basic Approach to Executive Decision Making*, American Management Association, 1978.

Parsons, T. and others, *Theories of Society; Foundations of Modern Sociological Theory*, The Free Press, 1961.

Perrow, C., *Normal Accidents: Living with High-Risk Technologies*, New York: Basic Books, 1984.

Piaget, J., *Introduction: The Place of the Sciences of Man in the System of Sciences*, UNESCO, 1970.

Pierce, J.R., *Symbols, Signals, and Noise: The Nature and Process of Communication*, Harper & Bro., 1961. (*An Introduction to Information Theory; Symbols, Signals and Noise*, Dover, 1980)

Piketty, T., *Capital in the Twenty-First Century*, Harvard University Press, 2014. (*Le Capital au XXIe siècle*)

Polanyi, M., *Personal Knowledge*, The University of Chicago Press, 1958.

Polanyi, M., *The Tacit Dimension*, Routledge & Kegan Paul, 1966.

Portland in USA, *Available* at: http://www.portlandoregon.gov/. (*Accessed* May, 2014)

Rapoport, A., "The Systemic View of the World", *World Congress of the Systems Sciences,* International Society for the Systems Sciences, Toronto, 2000, p.23.

Prigogine, I. and Stengers, I., *Order Out of Chaos*, Bantam Books, 1984.

Reason, J., *Human Error*, New York: Cambridge University Press, 1990.

Reason, J., *Managing the Risks of Organizational Accidents*, Ashgate Publishing, 1977.

Reason, J., Parker, D. and Free, R., "Bending the Rules: The Varieties, Origins and Management of Safety Violations", *Leiden: Faculty of Social Sciences*, University of Leiden, 1994.

Reason, J., *Review.I. Overview.II. Theory,* Derby: British Railways Board, 1993.

Reason, J., *The Human Contribution: Unsafe Acts, Accidents, and Heroic Recoveries,* Ashgate Publishing, 2008.

Reason, J., "The Chernobyl errors", *Bulletin of the British Psychological Society*, Vol. 40, 1987, pp. 2201–2206.

Reason, J., "The dimensions of safety" in Patrick, J. (ed.) *Cognitive Science Approaches to Cognitive Control*, Third European Conference, Cardiff: University of Wales, 1991.

Reitman, W.R., *Cognition and Thought; An Information Processing Approach*, Wiley, 1966.

Roethlisberger, F.J. and Dickson, W.J., *Management and the Worker*, Harvard University Press, 1939.

Rumelhart, D.E., *Introduction to Human Information Processing*, Wiley & Son, 1977.

Russell, P., *The Global Brain: Speculations on the Evolutionary Leap to Planetary Consciousness*, Curtis, 1982.

Sachs, J.D., *The End of Poverty: How We Can Make It Happen in Our Lifetime*, The Wylie Agency, 2005.

Saitō, H., "Fukushima genpatsu wa jishin ni taerareru no ka", in *Genpatsu wa jishin ni taerareru ka* [*Can Nuclear Plants Withstand an Earthquake?*], Citizen's Nuclear Information Center, 2008, pp. 29–33.

Sandel, M.J., *Justice: What's the Right Thing to do*, Farrar, Straus and Giroux, 2009.

Sandel, M.J., *What Money Can't Buy: The Moral Limits of Markets*, Farrar, Straus and Giroux, 2012.

Sandel, M.J., *Let's Talk about Justice in the World and a Special Lecture in Tohoku University*, Hayakawa Publishing, 2013.

Schein, E.H., *Career Dynamics: Matching Individual and Organizational Needs*, Addison–Wesley Longman, Incorporated, 1978.

Schein, E.H., *Organizational Culture and Leadership*, Jossey-Bass Publishers, 1985.

Schein, E.H., *Organizational Psychology*, 3rd edition, Prentice-Hall, 1980.

Schoderbek, C.G., Schoderbek, P.P. and Kefalas, A.G., *Management Systems*, Business Publications, 1980.

Schredinger, E., *Mind and Matter*, Cambridge, 1958.

Schredinger, E., *What is Life?*, Cambridge University Press, 1944.

Shannon, C.E., "A Symbolic Analysis of Relay and Switching Circuits", *TAIEE*, 1938, No.57, pp.1–11.

Shiozawa, Y., *Fukuzatsu-kei Keizaigaku Nyūmon* (Economies of Complex Systems), Seisansei Shuppan, 1997.

Silver, L.M., *Remaking Eden: How Genetic Engineering and Cloning will Transform the American Family*, Avon Books, 1997.

Simon, H.A. and Langley, P., "The Central Role of Learning in Cognition", *The Models of Thought*, Yale University, 1989, pp.102–115.

Simon, H.A. and Newell, A., *Human Problem Solving*, Prentice-Hall, 1972.

Simon, H.A. and Prietura, M.J., "The Experts in Your Midst", *Harvard Business Review*, Jan.-Feb.1989, pp.120–124.

Simon, H.A., *Administrative Behavior: A Study of Decision–Making Processes in Administrative Organizations*, Macmillan Publishing, 1945. (3rd edition, The Free Press, 1976)

Simon, H.A., *Models of Thought, Vol.3*, Yale University, 1979, pp.30–38.

Simon, H.A., Newell, A. and Shaw, J.C., "Elements of a Theory of Human Problem Solving", *Psychological Review*, 1958, Vol.65, pp.151–166.

Simon, H.A., *Reason in Human Affairs*, Stanford, 1983.

Simon, H.A., "Rational Choice and the Structure of the Environment", *Psychological Review*, Vol.63, 1956, pp.129–138.

Simon, H.A., "Simulation of Cognitive Processes; A Report on the Summer Research Training Institute", *Social Science Research Council*, 1953, Vol.12, pp.37–40.

Simon. H.A., *The Science of the Artificial*, MIT Press, 1985.

Sims, H.P., Gioia, D.A. and associates (eds.), *The Thinking Organization; Dynamics of Organizational Social Cognition*, Jossey-Bass, 1986.

Singh, J.V. and Baum, J.C., *Evolutionary Dynamics of Organizations*, Oxford University Press, 1994.

Skyrme, D., *Capitalizing on Knowledge: from e-business to k-business*, Elsevier plc group, 2001.

Skyrme, D., *Knowledge Networking: Creating the Collaborative Enterprise*, Elsevier plc group, 1999.

Smith, L. C., *The World in 2050: Four Forces Shaping Civilization's Northern Future*, Smith, L. C. 2010.

Smolka, A.P., "World map of natural hazards", Trans. Royal Society, A 2006; 364: 2147–2165.

Suzuki, D., *Inventing the Future*, Stoddart, 1989.

Suzuki, H. and Yamaguchi, T., *JR nishinihon no taizai: Hattori untenshijisatsujiken to Amagasakidassenjiko* [*The Mortal Sin of JR West*], Satsuki Publishing, 2007, p. 67.

Swanson, G.A., "Management Observation and Communication Theory", *Proceedings of the 44th World Congress of the ISSS*, International Society for the Systems Sciences, 2000, p. 121.

Swanson, G.E., *Social Change*, Scott Foreman and Company, 1971.

Taniguchi, T., *Sengo-nihon no kigyou-syakai to keiei-shisou*, Bunshindo, 2007.

Taylor, F. W., "Shop Management", in *Scientific Management*, Harper & Brothers, 1903, pp. 30–34.

Taylor, F.W., "The Principles of Scientific Management", in *Scientific Management*, Harper, 1911.

Taylor, F.W., "Taylor's Testimony before the Special House Committee", in *Scientific Management*, 1947.

The AIDS Scandal, Iwanami Booklet. *See also* (http://www.t3.rim.or.jp/~aids/yakugai2.html).
The ASAHI SHIMBUN, Doubt cast on Abe's assurance to IOC about Fukushima leaks, 10th Sep. 2013. *Available* at: (https://ajw.asahi.com/article/0311disaster/fukushima/AJ201309100071. (*Accessed* March, 2014)
The National Diet of Japan, *The official report of The Fukushima Nuclear Accident Independent Investigation Commission,* 2012. (*Available* at: http://warp.da.ndl.go.jp/info:ndljp/pid/3856371/naiic.go.jp/en/report/)
The World Bank, *Natural Disaster, Non-natural Disaster*, 2011, p. 30. (*Available* at: https://www.gfdrr.org/nhud-home)
Thom, R., Zeeman, E.C., Ujiki, S. and Sawa, T., *Keitai to kouzō—katasutrofu no riron* [*Catastrophe Theory*], Misuzu Shobō, 1977.
Toffler, A., *The Third Wave*, Morrow & Co., 1980.
Toffler, A., *Future Shock*, Morrow & Co., 1985.
Tokyo Shimbun, 23 August 2013.
Torgersen, P.E., *A Concept of Organization*, American Book, 1969.
Troncale, L.R., "Comparing Systems Pathology Treatments Across Systems Processes Theory", *Program and Abstracts of the 55th World Congress of ISSS,* International Society for the Systems Sciences, Hull UK, 2011, p. 115.
Troncale, L.R., "A Comprehensive Introduction to the Systems Sciences", *World Congress of the Systems Sciences*, International Society for the Systems Sciences, Toronto, 2000, p. 19.
Tsukio, Y., *Journeys beyond Time and Space: The Wisdom of the Indigenous Peoples*, Asahi Eco Books, 2012.
Tsuyumoto, I., *Dioxin*, Natsume, 2000.
Tung, R.L., "Human Resource Challenges and Opportunities to Outward Investment from Emerging Markets: The Case of China", *Proceedings of the 8th World Congress of IFSAM*, International Federation of Scholarly Associations of Management, Berlin, 2006, p.19.
Tung, R.L., "Comparative Perspectives on Expatriate Development and Appraisal", *Program and Abstracts of the 12th World Congress of IFSAM,* International Federation of Scholarly Associations of Management, Tokyo, 2006, p.19.
Tunstall, S.M., Johnson, C. and Penning-Rowsell, E. *Flood Hazard Management in England and Wales: From Land Drainage to Flood Risk Management,* World Congress on Natural Disaster Mitigation, New Delhi India, 19–21 February 2004.
Turner, B., *Man-made Disaster,* London: Wykeham, 1978.
Tönnies, F., *Gemeinschaft und Gesellschaft*, Leipzig: Fues's Verlag, 1887.
UNAIDS, *Global Report: UNAIDS Report on the Global AIDS Epidemic 2012*, pp. 6–8.
Ulrich, H. and Probst, G.J.B. (eds.), *Self-Organization and Management of Social Systems*, Springer-Verlag, 1984.
Vester, F., *Das Kybernetische Zeitalter*, Fischer, 1974.
WHO, World Health Organization, *Department of Control of Neglected Tropical Diseases*, 2012.
WWF, *The Energy Report —100 % Renewable Energy by 2050*, released November 10, 2011.
Watsuji, T., *Ningen no Gaku toshite no Rinrigaku* [Human Learning as Ethics], Iwanami Shoten, 1934.
Watsuji, T., *Fuudo: Ningengaku-teki Kousatu*, Iwanami, 1979.
Watson, L., *Lifetide: A Biology of the unconscious*, Hodder and Stoughton, 1979.
Weber, M., *Soziologische Grundbegriffe,* Verlag von J.C.B. Mohr (Paul Siebeck), 1921–22.
Weick, K.E. and Bougon, M.G., "Organizations as Cognitive Maps; Charting Ways to Success and Failure", Sims, H.P., Gioia, D.A., and Associates, *The Thinking Organization; Dynamics of Organizational Social Cognition*, Jossey-Bass Publishers, 1986, pp.102–134.
Weick, K.E., *The Social Psychology of Organizing, second edition*, The McGraw-Hill Companies, Inc., 1979.
Weick, K.E., "Organizational culture as a source of high reliability", *California Management Review*, 29, 1987, pp. 112–27.

Weick, K.E., "Sources of Order in Under-organized systems", Lincoln, Y.S. (eds), *Organizational Theory and Inquiry*, Sage Publications, 1985.

Weick, K.E., "The vulnerable system: an analysis of the Tenerife air disaster", Frost P.J. et al. (eds), *Reframing Organizational Culture*, Sage Publications, 1991.

Weizsacker, E.U. von, Hargroves, K., and Smith, M., *FAKTOR FUNF: Die Formel fur nachhaltiges Wachstum*, The Natural Edge Project, 2009.

West Japan Railway Company: labor union, June 1, 2005.

Whitehead, A.N., *Process and Reality: Science and the Modern World*, Macmillan, 1926.

Whitehead, A.N., *Symbolism Its Meaning and Effect*, 1927.

Wiener, N., *Cybernetics; or Control and Communication in the Animal and the Machine*, MIT Press, 1948.

Wiener, N., *The Human Use of Human Beings: Cybernetics and Society*, Houghton Mifflin, 1950.

Wilby J., Neal, R. and Bell, S., "Emergent Disaster Response during the June 2007 Floods in Kingston upon Hull, UK", *Journal of Flood Risk Management*, Wiley Online, 2011.

Williamson, O.E., *Organization Theory: Chester Barnard to the Present and Beyond*, Oxford University, 1990.

Wisner, B., Blaikie, P., Cannon, T. and Davis, I., *At Risk: Natural Hazards, People's Vulnerability, and Disasters (2nd Edition)*, Routledge, 2004.

Wolf, W.B. and Iino, H., *Philospphy for Managers: Selected Papers of Chester I. Barnard*, Bunshindo, 1986.

Yablokov, A.V., Nesterenko, V.B., Nesterenko, A.V. and Preobrazhenskaya N.E., *Chernobyl: Consequences of the Catastrophe for People and the Environment*, the New York Academy of Sciences, 2009.

Yamamoto, Y., *Keieigaku Honshitsuron* [*Essential Theory of Management*], Moriyama Shoten, 1961.

Yamamoto, Y. and Kato, K., *Keieigaku Genron*, Bushindo, 1982.

Yamamoto, Y. and Kato, K., *Keiei Hattenron*, Bushindo, 1997.

Yergin, D., *The Quest*, William Morris, 2011.

Yolles, M. and Iles, P., "Learning through Viable Knowledge Creation", *Proceedings of the 44th of the Systems Sciences*, International Society for the Systems Sciences, 2000, p.132.

Yoshida, T., *Jikososikisei no Jouhokagaku*, Shinyousha, 1990.

Yoshihara, M., *Keieigaku no Shikigen wo hiraita Sisouka-tati*, Bunshindo, 2006.

Zedtwitz, M. von., "A Four-layer Model for Studying Organizational Structure", *Academy of Management Best Papers Proceedings*, Seattle, 2003.

Index

S. Atsuji, *Unsafety*, Translational Systems Sciences 7,
DOI 10.1007/978-4-431-55924-5

Kolibri 360 GmbH
Ferdinand-Jühlke-Straße 7
99095 Erfurt, Deutschland
produktsicherheit@kolibri360.de